Atomic, Molecular, and Optical Science

An Investment in the Future

Panel on the Future of Atomic, Molecular, and Optical Sciences
Committee on Atomic, Molecular, and Optical Sciences
Board on Physics and Astronomy
Commission on Physical Sciences, Mathematics, and Applications
National Research Council

NATIONAL ACADEMY PRESS
Washington, D.C. 1994

NOTICE: The project that is the subject of this report was approved by the Governing Board of the National Research Council, whose members are drawn from the councils of the National Academy of Sciences, the National Academy of Engineering, and the Institute of Medicine. The members of the committee responsible for the report were chosen for their special competences and with regard for appropriate balance.

This report has been reviewed by a group other than the authors according to procedures approved by a Report Review Committee consisting of members of the National Academy of Sciences, the National Academy of Engineering, and the Institute of Medicine.

This project was supported by the Department of Energy under Grant No. DE-FG05-85ER 13326, the National Aeronautics and Space Administration under Grant No. NAGW-3282, the Air Force Office of Scientific Research and the National Science Foundation under Grant No. PHY-9100088, and the National Science Foundation under Grant Nos. PHY-8921799 and PHY-9222966. Partial support for this project was provided by the Basic Science Fund of the National Academy of Sciences, whose contributors include AT&T Bell Laboratories, Atlantic Richfield Foundation, BP America, Inc., Dow Chemical Company, E.I. du Pont de Nemours and Company, IBM Corporation, Merck and Company, Inc., Monsanto Company, and Shell Oil Companies Foundation.

Library of Congress Catalog Card No. 94-65036
International Standard Book No. 0-309-05032-4

Additional copies of this report are available from: National Academy Press 2101 Constitution Avenue, NW Box 285 Washington, DC 20055 800-624-6242 202-334-3313 (in the Washington Metropolitan Area)

B-285

Cover: Scanning electron microscopy with polarization analysis (SEMPA) image of magnetic domains in a patterned permalloy memory array. (Courtesy of J. Unguris, D.T. Pierce, and R.J. Celotta, National Institute of Standards and Technology.)

Copyright 1994 by the National Academy of Sciences. All rights reserved.

Printed in the United States of America

PANEL ON THE FUTURE OF ATOMIC, MOLECULAR, AND OPTICAL SCIENCES

GORDON H. DUNN, National Institute of Standards and Technology, *Chair*
LLOYD ARMSTRONG, Jr., University of Southern California
LOUIS E. BRUS, AT&T Bell Laboratories
SYLVIA T. CEYER, Massachusetts Institute of Technology
F. FLEMING CRIM, University of Wisconsin
ALEXANDER DALGARNO, Harvard-Smithsonian Center for Astrophysics
F. BARRY DUNNING, Rice University
ELSA M. GARMIRE, University of Southern California
PAUL L. KELLEY, Tufts University
DANIEL J. LARSON, University of Virginia
PAUL LIAO, Bellcore
STEPHEN R. LUNDEEN, Colorado State University
PETER W. MILONNI, Los Alamos National Laboratory
RICHARD C. POWELL, University of Arizona

Former Members of the Panel Who Were Active During the Study

NEAL LANE, Rice University, *Chair* (resigned July 1993 to become director of the National Science Foundation)
THOMAS J. McILRATH, University of Maryland (resigned September 1993 to become AMO program director at the National Science Foundation)
RONALD D. TAYLOR, Senior Program Officer

COMMITTEE ON ATOMIC, MOLECULAR, AND OPTICAL SCIENCES

DANIEL J. LARSON, University of Virginia, *Chair*
WILLIAM STWALLEY, University of Connecticut, *Vice Chair*
HOWARD C. BRYANT, University of New Mexico
SYLVIA T. CEYER, Massachusetts Institute of Technology
STEVEN CHU, Stanford University
DANIEL GRISHCHOWSKY, IBM T.J. Watson Research Center
WENDELL T. HILL III, University of Maryland
SIU AU LEE, Colorado State University
C. WILLIAM McCURDY, Lawrence Livermore National Laboratory
RONALD E. OLSON, University of Missouri
YUEN-RON SHEN, University of California, Berkeley
RICHART E. SLUSHER, AT&T Bell Laboratories
DAVID J. WINELAND, National Institute of Standards and Technology

Former Members of the Committee Who Were Active During Formation of the Panel

GORDON H. DUNN, National Institute of Standards and Technology
ANDREW W. HAZI, Lawrence Livermore National Laboratory
WILLIAM KLEMPERER, Harvard University
DONALD H. LEVY, University of Chicago
RONALD PHANEUF, University of Nevada, Reno
RONALD D. TAYLOR, Senior Program Officer

BOARD ON PHYSICS AND ASTRONOMY

DAVID N. SCHRAMM, University of Chicago, *Chair*
LLOYD ARMSTRONG, Jr., University of Southern California
DAVID H. AUSTON, Columbia University
DAVID E. BALDWIN, Lawrence Livermore National Laboratory
WILLIAM F. BRINKMAN, AT&T Bell Laboratories
PRAVEEN CHAUDHARI, IBM T.J. Watson Research Center
FRANK DRAKE, University of California, Santa Cruz
ROBERT C. DYNES, University of California, San Diego
HANS FRAUENFELDER, Los Alamos National Laboratory
JEROME I. FRIEDMAN, Massachusetts Institute of Technology
MARTHA P. HAYNES, Cornell University
GILLIAN KNAPP, Princeton University
ALBERT NARATH, Sandia National Laboratories
GEORGE W. PARSHALL, E.I. du Pont de Nemours & Company, Incorporated (retired)
JOSEPH M. PROUD, GTE Corporation (retired)
JOHANNA STACHEL, State University of New York at Stony Brook
DAVID WILKINSON, Princeton University
SIDNEY WOLFF, National Optical Astronomy Observatories
DONALD C. SHAPERO, Director
ROBERT L. RIEMER, Associate Director
RONALD D. TAYLOR, Senior Program Officer
TIMOTHY M. SNEAD, Administrative Associate
MARY RIENDEAU, Administrative Assistant
SUZANNE BOWEN, Program Assistant

COMMISSION ON PHYSICAL SCIENCES, MATHEMATICS, AND APPLICATIONS

RICHARD N. ZARE, Stanford University, *Chair*
RICHARD S. NICHOLSON, American Association for the Advancement of Science, *Vice Chair*
STEPHEN L. ADLER, Institute for Advanced Study, Princeton
JOHN A. ARMSTRONG, IBM Corporation (retired)
SYLVIA T. CEYER, Massachusetts Institute of Technology
AVNER FRIEDMAN, University of Minnesota
SUSAN L. GRAHAM, University of California, Berkeley
ROBERT J. HERMANN, United Technologies Corporation
HANS MARK, University of Texas at Austin
CLAIRE E. MAX, Lawrence Livermore National Laboratory
CHRISTOPHER F. McKEE, University of California, Berkeley
JAMES W. MITCHELL, AT&T Bell Laboratories
JEROME SACKS, National Institute of Statistical Sciences
A. RICHARD SEEBASS III, University of Colorado
CHARLES P. SLICHTER, University of Illinois at Urbana-Champaign
ALVIN W. TRIVELPIECE, Oak Ridge National Laboratory
NORMAN METZGER, Executive Director

Preface

In response to requests by several federal agencies, the Committee on Atomic, Molecular, and Optical Sciences (CAMOS) of the National Research Council's Board on Physics and Astronomy (BPA) proposed a study of atomic, molecular, and optical (AMO) science to be conducted by a panel chosen for that purpose. The specific charge to the Panel on the Future of Atomic, Molecular, and Optical Sciences (FAMOS) was to conduct an assessment of atomic, molecular, and optical science in the United States that reflects the opinions of the AMO community at large and addresses the following:

1. Determines manpower, instrumentation, facility, and funding requirements not only in the context of the intellectual challenges of AMO science, but also in the context of national needs such as (a) science education; (b) defense, energy, space, and environmental applications; (c) industrial and technological competitiveness; and (d) appropriate aspects of human health and welfare;
2. Seeks to identify scientific forefronts, technological opportunities, and windows of future opportunity;
3. Seeks to establish sets of research and educational priorities from various perspectives;
4. Sets forth goals and planning scenarios that reflect these research and educational priorities;
5. Develops long-range strategies that will best meet the goals set forth;
6. Assesses the institutional infrastructure in which AMO science is conducted and identifies changes that would improve its constituent research and educational efforts;

7. Provides a comparison of AMO science in the United States with its counterpart in other industrialized nations; and
8. Reviews the scientific advances made during the last decade.

With financial support from the U.S. Department of Energy, the National Science Foundation, the National Aeronautics and Space Administration, the Air Force Office of Scientific Research, and the National Research Council, the panel was assembled and has worked within the resources provided to meet the charge. This report is the product of the study.

AMO science centers on phenomena and processes involving the common building blocks of our world, that is, atoms, ions, molecules, and light, at energies characteristic of our everyday experience. Therefore, the new discoveries and inventions and the basic scientific knowledge provided by AMO science find immediate and widespread application in many areas of national importance. AMO science supports many areas of science, engineering, and technology and contributes significantly to the nation's highest priorities, including those related to basic scientific knowledge; education and human resources; industrial technology, manufacturing, and processing; information technology, high-performance computing, and communications; energy; global change; defense; health and medical technology; space technology; and transportation. Given the extraordinary diversity of the field and the broad impact of the science, the panel found it necessary in this assessment to adopt a focused definition of the field. However, the boundaries are occasionally crossed to better inform the reader of the applications and impacts of AMO science.

Briefly, atomic science encompasses the study of atoms and their ions, including their structure and properties; optical interactions; and collisions and interactions with electrons, external fields, and solids and surfaces. It is the test bed for the most precise tests of the fundamental physical laws of nature. Molecular science is defined here as the study of molecules, clusters, and molecular ions, including their structure and properties, optical interactions, collisions, and interactions with electrons, external fields, solids, and surfaces. Optical science here includes only those areas that are closely related to the laser, one of the key technological advances of this century, and to its applications. Thus, the panel has had to exclude many important areas of optical science and engineering, including vision, imaging science, atmospheric optics, and binary optics, and it hopes that these will be the subject of a future National Research Council assessment that will complement the present one.

Because AMO science provides data, understanding, instrumentation, and technologies that are essential in many other fields of science and engineering and in many applications, AMO science facilitates meeting a number of the nation's goals and needs. This report examines at some length this "enabling" aspect of the field.

At the organizational meeting in October 1991, the panel structured itself into subpanels that were disciplinary (atomic, molecular, and optical) and cross-disciplinary (technology impacts, education and human resources, interfaces with other sciences, resources and research infrastructure, and data) in nature. Most of the early work of the panel was carried out through these subpanels.

In an effort to gather the most current information from as broad a base as possible, letters asking for written input were sent to approximately 900 scientists in the field, to officers of the societies and divisions of societies that are closely allied to the field, and to heads of 11 major national laboratories. A number of industrial leaders were contacted. Town meetings were held at the International Quantum Electronics Conference (1992), the joint meeting of the Optical Society of America and the Interdisciplinary Laser Science Conference (1992), and the 1992 and 1993 annual meetings of the Division of Atomic, Molecular, and Optical Physics of the American Physical Society. A questionnaire was designed to solicit specific information from the scientists themselves that could be used to address the charge to the panel. The questionnaire was mailed to nearly 20,000 scientists who were members of the appropriate divisions, sections, interest groups, and so on, of the American Physical Society, the Optical Society of America, the American Chemical Society, the Institute of Electrical and Electronics Engineers, the Society of Photographic and Instrumentation Engineers, and the Materials Research Society.

A major challenge of the study has been that of assessing priorities, as set forth in the third item of the charge to the panel. AMO science affects national priorities in many different ways. It does not depend in its major thrusts on the construction and operation of large facilities. Because of these factors, a single linearized list of detailed technical priorities would not be meaningful. The panel, however, has arrived at three general priorities for AMO science in the immediate future.

The panel has been impressed by the important contributions to national needs that have been made and continue to be made by AMO science. These contributions range from the most fundamental levels of discovery and invention to applications critical to the nation's technological infrastructure and to its national economic productivity, competitive position, and security. The U.S. research investment in AMO science has yielded enormous returns both economically and in terms of scientific knowledge. Expecting this trend to continue, the panel views AMO science as the subtitle of the report indicates—*An Investment in the Future*.

THE NATIONAL ACADEMIES

National Academy of Sciences
National Academy of Engineering
Institute of Medicine
National Research Council

The **National Academy of Sciences** is a private, nonprofit, self-perpetuating society of distinguished scholars engaged in scientific and engineering research, dedicated to the furtherance of science and technology and to their use for the general welfare. Upon the authority of the charter granted to it by the Congress in 1863, the Academy has a mandate that requires it to advise the federal government on scientific and technical matters. Dr. Bruce M. Alberts is president of the National Academy of Sciences.

The **National Academy of Engineering** was established in 1964, under the charter of the National Academy of Sciences, as a parallel organization of outstanding engineers. It is autonomous in its administration and in the selection of its members, sharing with the National Academy of Sciences the responsibility for advising the federal government. The National Academy of Engineering also sponsors engineering programs aimed at meeting national needs, encourages education and research, and recognizes the superior achievements of engineers. Dr. Wm. A. Wulf is president of the National Academy of Engineering.

The **Institute of Medicine** was established in 1970 by the National Academy of Sciences to secure the services of eminent members of appropriate professions in the examination of policy matters pertaining to the health of the public. The Institute acts under the responsibility given to the National Academy of Sciences by its congressional charter to be an adviser to the federal government and, upon its own initiative, to identify issues of medical care, research, and education. Dr. Harvey V. Fineberg is president of the Institute of Medicine.

The **National Research Council** was organized by the National Academy of Sciences in 1916 to associate the broad community of science and technology with the Academy's purposes of furthering knowledge and advising the federal government. Functioning in accordance with general policies determined by the Academy, the Council has become the principal operating agency of both the National Academy of Sciences and the National Academy of Engineering in providing services to the government, the public, and the scientific and engineering communities. The Council is administered jointly by both Academies and the Institute of Medicine. Dr. Bruce M. Alberts and Dr. Wm. A. Wulf are chair and vice chair, respectively, of the National Research Council.

www.national-academies.org

Acknowledgments

The FAMOS panel acknowledges with thanks the thousands of workers in the field who took the time to respond to the questionnaire, attended the "town meetings," or otherwise provided input. Gratitude is also expressed to the more than 100 leaders and workers in the field who wrote detailed and thoughtful letters that helped guide the panel.

Agencies of the U.S. government were particularly cooperative in providing data and information for various parts of the study, and for this the panel especially acknowledges with thanks M. Berman (Air Force Office of Scientific Research), K. Gebbie (National Institute of Standards and Technology), W. Kalkofen (National Aeronautics and Space Administration), R. Kelley (Air Force Office of Scientific Research), A. Laufer (Department of Energy), J. Martinez (Department of Energy), R. McKnight (Department of Energy), P. Reynolds (Office of Naval Research), B. Schneider (National Science Foundation), D. Skatrud (Army Research Office), and J. Weiner (National Science Foundation).

A number of people were called upon to write or to critique portions of various drafts of the report, to supply figures for the report, or to give help and advice in other key ways. For this the panel thanks W. Anderson, N. Bardsley, H.G. Berry, J. Bowman, J. Burke, H. Carmichael, R. Celotta, S. Chu, J. Cohen, J. Dehmer, R. Deslattes, T. Deutsch, D. Dietrich, K. Dumas, G. Fisk, A. Gallagher, C. Gardner, T. George, H. Gould, N. Halas, B. Heckel, E. Heller, E. Herbst, E. Hinds, C. Howard, R. Hulet, S. Hurst, M. Inokuti, R. Jacobs, R. Keller, H.J. Kimble, N. Kurnit, S. Lamoreaux, S. Leone, R. LeSar, C.C. Lin, Chinlon Lin, J. Macek, G. Maggiora, L. Maleki, J. Margrave, J. McElroy, P. Meystre, A. Mills, A. Miziolek, P. Mohr, G. Pettit, A. Phelps, C. Rhodes, A. Shimony, J. Simons,

R. Smalley, B. Sundaram, A. Temkin, J. Tully, J. Weisheit, C. Wieman, L. Wilets, D. Wineland, and S. Younger.

Special appreciation is accorded Ronald Phaneuf, who did the statistical analysis of the questionnaire responses; to Steve Smith, who conducted a detailed analysis of support for the field; and to Kenneth Smith, who provided input for the economic impact analysis. Special thanks are also due to Christine Dunning, who provided the graphical interpretations of the questionnaire results, to Jennifer Overton and the staff of the Rice Quantum Institute for mailing the questionnaires and for data entry, and to Norma Cowley and Rita Mack for typing drafts of the questionnaire and report.

Finally, the panel gratefully acknowledges the work and support of the National Research Council and associated staff. Susan Mitchell from the Office of Scientific and Engineering Personnel provided advice in constructing and designing the survey as well as evaluating the results. Roseanne Price edited the final report. Suzanne Bowen from the Board on Physics and Astronomy provided general program assistance to the panel and prepared the final manuscript. Paul Uhlir of the Commission on Physical Sciences, Mathematics, and Applications worked with the panel as part of a Commission effort to explore objective, quantitative measures of the health of fields of science. This collaboration yielded many dividends for the panel's work. Ronald D. Taylor of the professional staff of the Board on Physics and Astronomy served as the study director throughout the project and provided valuable guidance on how to present the field in the most effective manner.

The idea for the study arose in the course of the Board on Physics and Astronomy's review of the topics covered by *Physics Through the 1990s* (the Brinkman Report; National Academy Press, Washington, D.C., 1986). After the Board raised the question of an update of the AMO volume of the Brinkman report, Steve Smith, then at the National Science Foundation, encouraged the Committee on Atomic, Molecular, and Optical Sciences to respond. The Committee on Atomic, Molecular, and Optical Sciences then developed detailed plans for the study. The panel thanks the BPA, CAMOS, and Steve Smith for the inspiration to undertake the preparation of this report. Thanks are also extended to the BPA for its interest and support throughout the study.

Contents

Executive Summary   1

PART I
OVERVIEW   5
   Introduction   7
   A Basic and Enabling Science   8
   Benefits of AMO Science   9
   Highlights of Scientific Advances   12
   The Scope and Support of AMO Science: The Core Program   13
   Findings and Recommendations   15
     Findings   15
       Impact of AMO Science   15
       Character of the Field   16
       Areas of Concern   17
     Recommendations   18
       Recommendations on Priorities   18
       Recommendations for the First Priority   18
       General Recommendations   19

PART II
ATOMIC, MOLECULAR, AND OPTICAL SCIENCE: TODAY AND TOMORROW — 21

1 Case Studies in AMO Science — 23
   Lasers: From Basic Research to New Technologies and New Industries — 23
   Manipulating Atoms: New Technology for Today and Tomorrow — 28
     Laser Trapping and Cooling of Atoms and Ions: Particle Optics — 28
     Optical Tweezers and the Biosciences — 30
   Buckyballs and Carbon Nanotechnology: Surprising New Materials from Small Science — 32

2 Recent Major Advances and Opportunities in AMO Science and Applications to the Needs of Society — 36
   The Nation's Scientific Knowledge Base — 36
     Recent Discoveries and Future Opportunities in AMO Science — 37
       Fundamental Laws and Symmetries — 37
       Cavity Electrodynamics and Micromasers — 39
       Highly Perturbed Atoms in Intense Laser and Microwave Fields — 41
       Transient States of Atomic Systems and Collision Dynamics — 42
       New Insights into Molecular Dynamics — 44
       Clusters — 47
       Physics of Nonlinear Optics — 48
       Laser Cooling and Trapping — 50
       Interactions with Surfaces — 52
     Enabling Other Fields of Science — 54
       Astrophysics — 55
       Space Science — 58
       Atmospheric and Environmental Science — 59
       Plasma Physics — 61
       Exotic Atoms and Nuclear Physics — 63
       Surface and Condensed Matter Physics — 65
       Biosciences—Mapping the Human Genome — 67
   The Nation's Measurement Technology — 68
     Measurement Standards — 68
     Measurement and Instrumentation — 70
     AMO in Measurement and Sensing for Industry — 71
   The Nation's Technological Infrastructure and U.S. Economic Productivity, Competitive Position, and Security — 72
     Industrial Technology, Manufacturing, and Processing — 72
       Lasers in Manufacturing — 72
       Plasma Processing of Materials — 75
       Chemical Manufacturing — 77
     Information Technology, High-Performance Computing, and Communications — 78
       The Erbium-Doped Fiber-Optic Amplifier — 79
       Optical Data Storage — 81

	Energy	81
	Energy Production	82
	Efficient Use of Energy	87
	Global Change	88
	Defense	91
	Weapons Systems and Delivery	92
	Remote Sensing	93
	Countermeasures	94
	C^3—Communication, Command, and Control	94
	Health and Medical Technology	96
	Medicine	96
	Radiation and Health Physics	100
	Design of Bioactive Molecules (Pharmaceuticals)	101
	Space Technology	102
	Measurement and Sensing	102
	Spacecraft Navigation and Communication	103
	Transportation	105
	Aviation	105
	Ground Transportation	106
3	Education and Human Resources	108
	Science Education	108
	K-12 Education	109
	Undergraduate and Graduate Education	109
	Human Resources in AMO Science	111
	Present Situation	112
	PhD Production and Initial Employment	113
	Future Needs	114
4	Funding and Infrastructure for Research and Development in AMO Science	116
	Resources	116
	Federal Funding for Research in AMO Science	119
	National Science Foundation	119
	Department of Energy	121
	Department of Defense Research Offices	122
	National Aeronautics and Space Administration	123
	Total Funding from Federal Grants and Contracts	123
	Federal Laboratories	125
	National Institute of Standards and Technology	125
	Department of Energy Laboratories	125
	Department of Defense Laboratories	126
	Infrastructure and Facilities	126
	The Single Investigator	126
	Centers and Institutes	127
	User Facilities	128
	National Laboratories	130

		Other Infrastructure Issues	130
		Evolution of Subfields	130
		Theory	131
		Instrumentation	132
		Academic Culture	132
		Postdoctoral Associates/Researchers and PhD Employment	133
		Communication and Organization	133
5	Economic Impact of AMO Science		135
6	International Perspectives in AMO Science		141
	APPENDIXES		
A	Nobel Prizes Awarded in AMO Science Since 1964		147
B	Impact of AMO Science		155
C	Citation Analysis		159
D	Survey of AMO Scientists		163

Atomic, Molecular, and Optical Science

An Investment in the Future

Executive Summary

At this time of great change in the world, it is appropriate to assess the value of the nation's investment in basic science research and education. Today it is all the more important that science and the technology it produces contribute to solving society's problems and that the role of science and technology be understood. The challenge in developing a national science research and education strategy is to select and enhance the areas of basic research that constitute the wisest investment of the nation's resources for long-term as well as immediate benefits.

This report describes the results of a study commissioned by the National Research Council to assess the field of atomic, molecular, and optical (AMO) science "not only in the context of the intellectual challenges of AMO science, but also in the context of national needs."

AMO science focuses on the properties of the common building blocks of the world around us—namely, atoms, molecules, and light—and on phenomena that occur in the ranges of temperature and energy that are characteristic of daily human activities. It is both an intellectually stimulating basic science and a powerful "enabling" science that supports many other important areas of science and technology. It is a key component of several of the federal strategic Federal Coordinating Council for Science, Engineering, and Technology (FCCSET) initiatives. The enabling aspect of AMO science derives from efforts to control and manipulate atoms, molecules, charged particles, and light more precisely; to accurately measure and calculate their properties; and to invent new ways to generate light with specific properties.

Federal funding of grants and contracts for basic research in AMO science in the United States amounts to somewhat over $100M per year, provided primarily by the National Science Foundation, Department of Energy, and Department of Defense. If support provided by U.S. industry and through U.S. federal research and development laboratories is included, the total funding for AMO science in this country approaches $1B per year. The return on this investment is substantial. The panel estimates that AMO science is an important enabling factor in industries accounting for about 9% of the nation's GNP. Overall, the products of AMO science influence over 20% of the GNP (*U.S. Industrial Outlook 1992: Business Forecasts for 350 Industries*, International Trade Administration, U.S. Department of Commerce, U.S. Government Printing Office, Washington, D.C., 1992).

Despite its impact, AMO science is a "small" science. That is, it is dominated by the work of single investigators or small groups. This mode of research has proved to be an effective vehicle for creative and innovative science and has resulted in many notable discoveries. Students who graduate with backgrounds in AMO science acquire a broad range of knowledge and skills and are valuable contributors to many areas of science and technology.

The report highlights recent advances and discoveries in AMO science. Through examples, it illustrates many areas of application, including industrial technology, manufacturing, and processing; information technology, high-performance computing, and communications; energy; global change; defense; health and medical technology; space technology; and transportation.

Three case studies—"Lasers," "Manipulating Atoms," and "Buckyballs and Carbon Nanotechnology"—are included that show different stages in the transfer of basic scientific knowledge and technologies to direct applications and marketable products. The special role of AMO-science-based measurement techniques, instrumentation, and sensors in the nation's manufacturing industries is also discussed. Funding, research infrastructure, and education and human resources are also addressed.

The report identifies a number of concerns about the future vitality of the U.S. programs in AMO science and its applications. Substantial support for the field has come from defense programs within the Department of Defense and Department of Energy. Conversion of defense research and development funding to commercial objectives could result in serious erosion in the support of AMO science. Industrial research and development laboratories, which have supported substantial AMO science research activities, are reorganizing and cutting back. Demand for PhD graduates is down in all fields, a trend that is less of a problem for AMO graduates because of their broad training and practical skills. There is also concern within the AMO science community that the United States is losing ground to other nations in many areas of AMO science.

The recommendations presented in the report include guidelines to federal agencies on general priorities. Mechanisms to help mission agencies establish specific priorities are suggested. The report further recommends that the agencies emphasize the support of single investigators and small groups, rely on merit review for undirected as well as goal-directed basic research, and support an interagency advisory or coordinating committee to collect and disseminate information and provide guidance to government, industry, and the AMO science community.

The report is divided into two parts. Part I, "Overview," discusses the impact of the science and recent discoveries, and presents recommendations aimed at enhancing the value of the science to society. Part II, "Atomic, Molecular, and Optical Science: Today and Tomorrow," the main body of the report, amplifies and gives details of topics summarized in the overview.

The panel has arrived at three priorities for AMO science in the immediate future. First, the panel recommends a pattern of support that maintains and enhances responsiveness of AMO science to national needs by ensuring the healthy diversity of the field and the strength of the core research. Second, the panel recommends research into highly promising new technologies for the control and manipulation of atoms, molecules, charged particles, and light. Third, the panel recommends research into new and improved lasers and other advanced light sources. Specific priorities with regard to different national goals and needs —such as those related to industrial technology, manufacturing, and processing; information technology, high-performance computing, and communications; energy; global change; defense; health and medical technology; space technology; and transportation—can best be identified by agencies and advisory bodies addressing these specific goals and needs.

EXECUTIVE SUMMARY

Part I

Overview

This overview is divided into several sections. The context in which the report was prepared is described in the "Introduction." AMO science touches on the most basic aspects of the physical world. It also makes extensive contributions to societal needs and serves as the underpinning for a number of other areas of science. At a time when society is increasingly interested in understanding the benefits and returns of basic research, this field is able to give an impressive accounting of itself. The nature of the field is described in the next section, "A Basic and Enabling Science." Some of the fundamental questions addressed by AMO science are discussed, and its role in enabling other sciences, applications, and technology is explored. In "Benefits of AMO Science," the field's contributions to scientific and human resources, to economic productivity, to medical science, and to the nation's technological infrastructure are described. "Highlights of Scientific Advances" reviews some of the achievements of the science, including cooling, trapping, and manipulation of individual atoms and molecules and measurements of the course of chemical reactions as they occur. The U.S. research program in this field is described in the next section. Two priority areas identified by the panel are discussed: control and manipulation of atoms, molecules, charged particles, and light, and the development of new light sources. The case is made that support of the basic research program will continue to yield rich dividends in such areas as economic competitiveness, health care, energy, the environment, and others. The final section of the overview outlines the findings and recommendations of the study panel.

INTRODUCTION

Increasingly, society is asking for greater accountability from scientists and evidence of a return on its investment in scientific research. Assembled to assess the field of atomic, molecular, and optical (AMO) science in that context, the Panel on Future Opportunities in Atomic, Molecular, and Optical Sciences was charged, among other things, to review advances of the last decade; determine requirements of the field in the context of national needs such as those related to industrial and technological competitiveness, human health and welfare, environment, defense, energy, and education; establish research and educational priorities from various perspectives; and identify scientific forefronts, technological opportunities, and windows of future opportunity.

As demonstrated in this report, AMO science is a diverse field whose impact on national needs and priorities is substantial. The panel estimates that AMO science is an important enabling factor in industries accounting for about 9% of the nation's GNP. Overall, the products of AMO science influence over 20% of the GNP (*U.S. Industrial Outlook 1992: Business Forecasts for 350 Industries*, International Trade Administration, U.S. Department of Commerce, U.S. Government Printing Office, Washington, D.C., 1992).

Indeed, the field is so broad and intersects so many other fields of science and engineering that comprehensive coverage is impractical. For the purposes of this report, a circumscribed definition of the field has been adopted. Developing a general rule to limit coverage has not been easy, and occasionally the panel has put aside the rule to better inform the reader of the applications and impacts of AMO science.

Atomic science encompasses the study of atoms and their ions, including their structure and properties; optical interactions; and collisions and interactions with electrons, external fields, and solids and surfaces. It is the test bed for the most fundamental laws of science. Topics of interest include fundamental laws and symmetries; cavity electrodynamics; transient states of atomic systems and collision dynamics; highly perturbed atoms; cooling and trapping; atom interferometry; and interactions with surfaces.

Molecular science is also a diverse field that spans a broad range of research areas and applications, including most of chemistry and significant portions of biology. To narrow the focus for purposes of this report, the panel defines molecular science as the study of molecules, clusters, and molecular ions, including their structure and properties, optical interactions, collisions, and interactions with electrons, external fields, and solids and surfaces. The report emphasizes, in particular, molecular interactions at the quantum state resolved level; ultrafast phenomena in molecules; clusters and molecular aggregates; and interactions with surfaces. All areas can point to important accomplishments with broad impacts, and, characteristic of AMO science, they interact strongly with each other.

Optical science has become an integral part of many disciplines, ranging from biology to astronomy, and has found application in key economic segments from medicine to telecommunications. However, it is not possible to cover all the diverse aspects of the optical sciences in a report that also covers atomic and molecular science. Because one of the key advances in this century has been the invention of the laser, the panel limits this report to those optical science areas that are closely related to the laser and its applications. These areas have been the subject of the International Quantum Electronics Conferences since they began in 1967 and, more recently, the Quantum Electronics and Laser Science Conferences, which have been held yearly since 1989. Hence, this report focuses on the topics that are within the purview of these conferences and adopts a definition of optical science that includes only the following subjects: laser spectroscopy; nonlinear optical phenomena; quantum optics; optical interactions with condensed matter; ultrafast optics; and coherent light sources.

Although this definition provides a focus for this report and encompasses many of the exciting topics in this field, it does not include many equally important areas of optical science. In what might be called "classical optics," for example, advances in the design of lenses for lithography have been essential to improvements in integrated circuit technology. Indeed, optical instruments of all kinds are routinely used in virtually every aspect of modern life. Many other areas of optics, including vision, imaging science, atmospheric optics, and binary optics, are also not considered here. These areas would be more appropriately addressed in a broader study of optical science and engineering.

The focus of the study and this report is on basic, or in-depth, research, whether supported with a strategic objective in mind or simply because of its intellectual excitement. Some of the examples chosen by the panel include applied research and development in order to show the ultimate application of AMO science in a variety of areas and to demonstrate that basic research was a necessary precursor of these applications. Thus, making a precise distinction between basic and applied research in each case is not important.

The present study emphasizes advances and opportunities in AMO science and its applications to national needs; human resources and funding in the field; recommendations, including priorities; and the results of a survey of the AMO science community. The report also briefly discusses the infrastructure in which AMO science is conducted and makes limited comparisons with efforts in other countries.

A BASIC AND ENABLING SCIENCE

AMO science is simultaneously a basic and an enabling science that answers questions about the behavior of matter and energy in atomic and molecular systems that we can precisely probe, control, and manipulate. It focuses on the common building blocks of the world around us, that is, atoms, molecules, and light, and on phenomena that occur in the ranges of temperature and energy that are characteristic of daily human activities.

As a basic science, AMO research provides answers to fundamental questions about the physical world and accurate tests of basic physical theories such as quantum electrodynamics, quantum measurement, relativity, electroweak interaction, time reversal, and the invariance of the combined operations of charge conjugation, parity inversion, and time reversal (CPT).

In its enabling role, AMO science has, throughout its extensive history, contributed to the technological strength and knowledge base of the nation. In recent years, the field has continued to grow in excitement and activity, fueled by the discovery of new phenomena and the widespread practical application of the science, all of which have been facilitated by the development of new experimental, theoretical, and computational tools.

The rapid pace of new discoveries and developments in AMO science can be attributed to the continued invention and implementation of new techniques to control and manipulate atoms, molecules, and light and to generate light with well-defined characteristics. These, in turn, permit new measurements of natural phenomena. The theme of control, manipulation, and measurement that so well characterizes AMO science underscores the impact of the field, because these capabilities have important applications in all branches of science, engineering, and technology. The world's most accurate measurements occur in AMO science, because time and frequency, which are the most accurately measurable physical quantities, fall primarily in the AMO domain. The quest of AMO scientists for improved measurement techniques and accuracy has resulted in inventive new instrumentation, including new sources of light, and technologies that find application in areas ranging from industrial manufacturing, new materials, and processing to medicine and environmental monitoring.

BENEFITS OF AMO SCIENCE

The goal of the U.S. program in AMO science is to improve the nation's

- Scientific and human resources, by advancing basic scientific knowledge through invention or discovery of new technologies and through measurement and calculation of the properties of atoms, molecules, and light and by contributing to science education;
- Economic productivity, competitive position, and security, through research leading to new materials and processes, to new manufacturing, information, medical, and other technologies and methodologies, and to strategically important physical data on AMO systems; and
- Technological infrastructure, through the development and application of new instrumentation, new experimental and computational methods and systems, and expanded databases.

The societal benefits of AMO science can best be illustrated with a few examples. A more detailed discussion is given in Part II of this report, "Atomic, Molecular, and Optical Science: Today and Tomorrow."

Understanding global change and the impact of human activities on the environment requires reliable models of chemical and physical responses of the atmosphere to radiation from the sun that are based on AMO science and utilize the basic information about the properties and behavior of atoms, ions, and molecules as they interact with light and with one another in the atmosphere or hydrosphere. Laser techniques, including laser-radar (LIDAR), are employed for testing as well as providing empirical data for the models. Laser technology is used to monitor emissions, effluent, and toxic waste environments.

The commanding international currency is energy. AMO science is important to essentially every aspect of energy production and has long played a vital role in the development of advanced energy systems. Its importance will only increase as the nation moves toward more efficient and cleaner energy sources. AMO science is one of the keys to optimization of combustion, solar, fission, and fusion energy systems and lighting efficiency. Information on energy transfer processes involving collisions of electrons with ions, atoms, and molecules and collisions of ions, atoms, and molecules with one another and with surfaces is critical in many applications.

Weapons guidance, detection of intrusion by hostile weaponry, monitoring of possible poison gases, and design and analysis of the effects of weapons depend on input from AMO science, illustrating that the contributions of AMO science to national defense and security have been substantial and will continue to play a vital role as new optical technologies are introduced and as simulations of warfare scenarios become increasingly important. AMO science will be a major element in the sensing technology necessary for monitoring compliance with arms limitation agreements in an increasingly complex geopolitical scene.

Communications technology has, in the past two decades, been revolutionized through the use of fiber optics. Indeed, the vision of what lies ahead in this area is so expansive that one may say that the revolution has just begun. The improvement in the quality of optical fibers and the discovery, development, and application of the semiconductor diode laser have been a joint triumph of solid-state, AMO, and materials science. Almost invisible to the user, this technology allows the inexpensive delivery of information for government, commerce, industry, and academe over great distances, at high speeds, and in large volume. Future advances in AMO science and technology will allow even greater fiber bandwidth and flexibility of network connection, resulting in "on-demand" access to a vast, worldwide storehouse of information, including high-resolution video.

In computer technology, AMO science has had an impact through the optical storage of information on CD-ROMs and the computer-to-computer links provided by fiber optics. In the future, it is likely that optics, because of its speed, freedom from interference, and design advantages, will play an important role inside computers, by providing links from circuit board to circuit board and from chip to chip. In the commercial marketplace, CD laser music players and laser video disks as well as supermarket laser checkout devices have, in only a few years, become commonplace in our lives.

Advanced manufacturing methods depend on AMO science. The use of lasers to cut, weld, drill, mark, and trim materials is widespread today. Chemical and optical sensors are essential in process control, and measurements based in AMO science are essential to quality control. Knowledge and control of atomic and molecular processes occurring in plasma processing of materials used in electronic chip manufacture and in the aerospace, automotive, steel, biomedical, and toxic waste management industries help in the effectiveness and efficiency of those operations.

The world's measurement standards are primarily based in AMO science, and a large fraction of modern measurement methodologies and instruments originate in AMO science. Accurate and precisely interrelated measurements are essential to equity in trade, quality control in manufacture, access to the global marketplace, and the progress of science itself. Efforts toward providing better measurements and the benefits that follow are continuously progressing.

Medical science and technology have benefited from AMO science in a variety of ways. Molecular physics and chemistry have traditionally played an important role in the understanding of chemical bonding, biomolecular structures, and the dynamics of energy transfer in biological molecules. Supporting this role are tools provided by AMO science, such as excitation, Raman, and ultrafast (short time duration) spectroscopy. Ultrafast optical spectroscopy techniques developed in the last 20 years have begun to unravel primary photophysical events in biological systems. They have probed visual pigment isomerization, electron transfer at the photosynthetic reaction center, and ligand binding in hemoglobin. All of these measurements address questions at the heart of molecular biology. Looking ahead, it has been found recently that buckminster-fullerence (C_{60}) molecules, or "buckyballs," a recent discovery of AMO science, neutralize a large area of the HIV virus, thus introducing the possibility that this new particle may help in the battle against AIDS.

Medicine also has benefited from AMO science in terms of the application of image science and technology to visualization of the body and its diseases and the use of laser radiation as a tool to modify microscopic cells and macroscopic tissue. In the latter area, laser radiation has been used to clear coronary arteries and break up kidney and gall bladder stones, and for retinal welding, corneal sculpting, and photochemical release of oxygen from chromophores attached to cancer cells. When combined with optical fibers, lasers allow minimally invasive surgery with smaller incisions that result in less danger and quicker recovery. Laser radiation has several uses in clinical practice, while other promising uses require further research and clinical trials.

Technologies arising from AMO science promise to improve the safety, speed, and efficiency of transportation systems. In aviation, AMO science is contributing to the development of systems to detect wind shear/clear air turbulence and wake vortices. The Global Positioning System (GPS), which is based on atomic clocks, is an important practical application of AMO science. Developed originally for the military, inexpensive commercial receivers are now becoming readily available, allowing the technology to be used for commercial and recreational purposes.

HIGHLIGHTS OF SCIENTIFIC ADVANCES

The examples of direct benefits of AMO science to society described in the previous section are follow-ons of basic, in-depth research. The past decade represents a period in which the excitement accompanying fundamental advances in AMO science has been extraordinarily high, as it is today. Because of its diversity, AMO science has many "cutting edges" along which the science advances, as well as many interfaces with other fields of science, engineering, and applications. The list of Nobel Prizes given for work in AMO science, presented in Appendix A, helps illustrate the point. On the basis of past experience, one can expect that many of these scientific advances will one day lead to applications. In this section, brief descriptions are given of a few examples. Again, the reader is referred to Part II of this report for a more detailed discussion and other examples.

Atomic particles, trapped in various combinations of magneto-optical traps, have been cooled to temperatures of the order of one-millionth degree above absolute zero—perhaps the lowest temperature in the universe, and certainly the lowest ever achieved on Earth—allowing demonstration of atom interferometers.

AMO scientists have discovered new, exotic forms of matter, the best known of which is, perhaps, the "buckyball." This is a member of a class of new carbon compounds referred to as fullerenes, which also include carbon "nanotubes." Progress in this area is rapid, and new related materials, including high-temperature superconductors, are being actively pursued.

Real-time measurements of atomic motions within a molecule have now been made, thus allowing detailed study of chemical reactions and how energy is transferred from one part of a molecule to another.

Also discussed in Part II of this report are advances in cavity electrodynamics and micromasers, highly perturbed atoms, chaos in atoms and molecules, collisions and transient states of atomic systems, nonlinear optics, surface interactions, and other topics. The fact that AMO science is an enabling science to other fields—astrophysics, space science, atmospheric and environmental science, biosciences, plasma science, nuclear physics, and surface science—is also emphasized there, and examples are discussed.

Already, some of the more recent basic research results mentioned above are being vigorously pursued for their possible applications. For example:

- Atom trapping and cooling have been extended to make "optical tweezers" that have been developed into a commercial product and used in biological studies to manipulate bacteria, deoxyribonucleic acid (DNA), and other molecules of life. These tweezers will likely be invaluable in the Human Genome Project.
- Carbon nanotubes are being grown in longer and longer lengths, and speculation is that they may reach centimeter lengths. They could be the strongest fibers known, and the electrical and structural materials possibilities could be substantial.
- The ability to monitor chemical reactions as they occur and to selectively input energy to the reactants holds the promise of controlling chemical reactions and tailoring their outcome to achieve specific products.

These examples are not exhaustive, but they illustrate what has always been true about science, that basic research of an exploratory nature often opens up new applications of benefit to society in unexpected ways. At the same time, there are important areas of research (basic as well as applied) that are directed toward specific applications. For example, in the case of energy or the environment, it is necessary to have a large amount of information, including reliable data, on the properties of atoms, molecules, and charged particles and their interactions with one another and with light and other fields. To obtain such information requires a great deal of research, often requiring the development of new techniques for measurement and observation. Although this is basic research, it is directed toward the goal of dealing with the nation's energy and environmental needs. To cite an example in the area of medicine, further advances in laser surgery may require lasers of particular frequencies and characteristics, necessitating goal-directed research. However, it is vitally important to the field and its future applications to maintain a balance between goal-oriented research and exploratory basic research.

THE SCOPE AND SUPPORT OF AMO SCIENCE: THE CORE PROGRAM

The foregoing discussion and supporting information in Part II of this report make the case that AMO science produces knowledge and technologies that are of social and economic significance and that such benefits will continue in the future. Before proceeding to detailed findings and recommendations for the field, it is important to examine questions of appropriate size and scope of the U.S. program in AMO science, the current level of financial support for the field, and the way in which that support should be invested.

The years since World War II, particularly the post-Sputnik era, saw a large growth in scientific research including the AMO sciences. Large research universities grew larger, and the number of research universities increased and became more geographically dispersed. With this growth came an increase in the number of AMO scientists in universities and colleges; further, large numbers of AMO scientists became engaged in research and development in government and industrially supported laboratories. That growth stopped some time ago and in the case of industry has been reversed.

Because of the diversity of AMO science—with its many subfields and intersections with other fields of science and engineering—it is not possible to define, precisely, the size of the AMO science community or the total level of support of the core basic research program. However, the panel estimates that, currently, the U.S. research program in AMO science includes 6,000 to 7,000 PhD researchers. The grant and contract support from federal sources alone for basic research in universities, colleges, and selected federal laboratories exceeds $100M annually, but this is only a fraction of the total; if industrial support and funds provided through the federal research and development laboratories are included, the annual funding level is many times this amount and may approach $1B. The size of the field and total level of funding, in real terms, have not changed significantly in several years.

The U.S. core program of basic research in AMO science has developed to its present size and scope largely because of the diversity and importance of applications of AMO science in the areas described above. There is no reason to believe that the products of the field will be any less in demand in the future. Indeed, given the emphasis being placed on technology, maintaining a healthy core program in AMO science will be even more critical. The panel estimates that AMO science is an important enabling factor in products and services amounting to about 9% of the GNP, which suggests that the nation's investment in the field has been economically beneficial. The panel's top priority is maintaining and enhancing responsiveness of the field to national needs by ensuring a healthy, balanced core program of research and education in AMO science.

Within the core program, there will always be areas where modest increases in funding can yield enhanced returns. Identification of these areas is best made by the agencies, through merit review of individual proposals and new programs. However, the panel notes that advances in technology have frequently triggered major advances in the science and thus recommends two areas of priority: the control and manipulation of atoms, molecules, macroscopic particles, and light,[1] and the invention and development of new sources of light.[2] In both of these cases, new technologies as well as new knowledge are likely outcomes.

---

[1] "Control and manipulation" applies to internal states as well as energy, time, momentum, and position and includes the invention and development of methods to enable measurements of atoms, molecules, charged particles, and light at higher resolution.

[2] "Invention and development of new sources of light" includes new types of lasers and related optical technologies but is not limited to these systems. The emphasis here is on new ways to produce light with well-defined characteristics—wavelength, power, pulse width, coherence, and other properties.

In the areas of application of AMO science to national needs, it is understood that mission-oriented government agencies and industry will organize the science in terms of the applications that are envisioned. Basic science can and should be encompassed within such an organization to realize all the potential benefits of AMO science. Without the mission agencies' strong involvement in supporting this field, the U.S. core program would be much weaker, and many applications of the science would have been lost. The Department of Defense (DOD) and the Department of Energy (DOE) have understood the need for this undergirding support for AMO science and have traditionally supported strong research programs, the products of which have led to applications to their respective missions. As the nation's concerns shift from defense requirements toward civilian needs such as economic competitiveness, national infrastructures, health care, energy, and the environment, it must be recognized that the products of AMO science may be even more valuable to these new goals than they have been to defense.

FINDINGS AND RECOMMENDATIONS

During the course of this assessment of AMO science, the panel reached a number of conclusions concerning the vitality, character, and impact of the field and identified a number of concerns. These findings and concerns are presented here, together with the panel's recommendations and statement of priorities, which are intended to ensure that AMO science will continue to provide significant benefits to society. Further justification for these findings and recommendations can be found in Part II of the report.

FINDINGS

IMPACT OF AMO SCIENCE

- AMO science, a rapidly evolving basic science and a powerful "enabling" science, contributes to the fundamental knowledge base and supports important areas of science, engineering, technology, and applications.
- The nation's investments in AMO science research and education have yielded substantial economic benefits. The panel estimates that AMO science, through its applications to manufacturing, information technology and communications, semiconductors, and other commercial sectors, is an important enabling factor in industries accounting for about 9% of the nation's GNP. Overall, the products of AMO science influence over 20% of the GNP.
- AMO science is diverse, and the base of scientific knowledge, methods,

and technologies it provides plays a critical role in many areas of science and technology, including applications to industrial and information technology, energy and environment, health, space technology, defense, and transportation.
- AMO science has much to contribute to the federal strategic initiatives, including those related to advanced materials and processing; advanced manufacturing technology; global change research; high-performance computing and communications; science, mathematics, engineering, and technology education; and biotechnology research.
- Measurement techniques, sensors, and instrumentation based on AMO science are a central component of advanced manufacturing processes and contribute significantly to enhanced industrial output. They are also important to environmental monitoring, pollution control, and medical diagnostics and monitoring.
- Students educated and trained in AMO science acquire a broad range of knowledge and skills and are valuable contributors to many areas of science and technology. They are employed by industries that have contributed significantly to recent economic growth in the United States and that are likely to be important in sustaining its economic health.

CHARACTER OF THE FIELD

- AMO scientists use experimental, theoretical, and computational methods to study matter at the atomic level. Their activities involve the control and manipulation of atoms, molecules, charged particles, and light, the measurement and calculation of their properties, and the generation of light with well-defined characteristics with the overall objective of understanding the structure and dynamics of atoms, ions, molecules, and light and the nature of their interactions.
- AMO science is "small" science and is most often carried out by principal investigators and their co-workers in small groups, frequently in collaboration with other scientists and engineers. This scale of research has proved to be an excellent vehicle for creative and innovative science and has spawned notable achievements in the field. Clustering of small groups at "centers" and special facilities is sometimes necessary for interdisciplinary research and research needing special facilities that, for cost or other reasons, are difficult to duplicate.
- AMO science in universities and colleges and in government laboratories receives support from a range of federal agencies, reflecting the breadth and diversity of the field. Its advancement has been facilitated by the use of merit review to identify and fund the best research projects, resulting in a U.S. program that is, in many areas, the strongest in the world.
- AMO science is funded at substantial levels in federal research and development laboratories supported by DOE, DOD, the National Aeronautics and Space Administration (NASA), and the National Institute of Standards and Technology (NIST) and in industry. The AMO scientists in these laboratories, their

- knowledge and skills, and their experimental and computational facilities are a vital component of the national AMO science program.
- Several federal agencies have recognized the need to maintain a healthy AMO community and the importance of basic AMO research to mission objectives. This has resulted in a balanced AMO national program that has a tradition of innovative research, that has achieved a broad and impressive range of strategic goals, and that has trained many highly qualified young scientists.
- The panel found a strong belief among researchers familiar with foreign laboratories that the United States is falling behind in the quality of instrumentation in its academic research laboratories. Workers in the field indicated that updating of capital equipment continues to be a high priority.
- AMO science engages about 6,000 to 7,000 active PhD researchers. This number has remained essentially constant over the past decade, although there has been a redistribution of AMO scientists among areas of specialty, with many more scientists working in optical science than in the past. This shift is in part a response to industrial needs. The overall level of activity is adequate to sustain a strong and dynamic program, and there appear to be no compelling reasons for change in the immediate future.

AREAS OF CONCERN

- Substantial support for basic AMO scientific research has come from DOD and defense areas of DOE. Despite the demonstrated application of the fruits of AMO research in areas such as medicine, environment, transportation, and commerce, there is a danger that the shift in federal funding from defense could result in serious erosion of basic research in AMO science.
- Industrial and federally funded laboratories have been an important component of the U.S. research effort in AMO science. Several of these laboratories at this time are undergoing major reorganizations and reductions that could be seriously detrimental to the U.S. program in AMO science.
- Many important practical applications of AMO science in areas such as remote sensing, atmospheric science and fossil fuel combustion, plasma processing, and medical diagnostics require a database of accurate quantitative measurements and calculations of atomic, molecular, and optical properties. Support for such essential core work in AMO science can be negatively affected in times of limited funding by pressure to support research in more exotic areas.
- Although the production of new PhDs in AMO science has been approximately constant for a decade, demand is down as it is in other fields and, despite their broad training, many young scientists are unable to find permanent positions.
- Although the U.S. core program in AMO science is strong, there is concern in the community that the U.S. program is losing ground relative to those in Europe and Japan.

RECOMMENDATIONS

RECOMMENDATIONS ON PRIORITIES

- The **first priority** is to maintain and enhance responsiveness of AMO science to national needs by assuring the vitality and diversity of the U.S. core program in experimental, theoretical, and computational AMO science in academic institutions, industry, and government.

    History has shown that many advances in AMO science and in its applications have been triggered by the invention and development of new techniques, instrumentation, and technology, the most notable by the invention of the laser. The second and third priorities focus on this enabling aspect of the field.

- The **second priority** is to promote research that promises new technologies through the invention and development of techniques and instrumentation to better control and manipulate atoms, molecules, charged particles, and light for a broad range of applications and for furthering studies of interactions at the atomic and molecular level.[3]

- The **third priority** is to promote research that promises new and improved lasers and other advanced sources of light for a broad range of applications and for furthering studies of the properties of light and its interaction with atoms and molecules.[4]

RECOMMENDATIONS FOR THE FIRST PRIORITY

To achieve the first priority, the panel recommends several actions.

- The panel recommends that balanced involvement of the field in both basic and strategic research be maintained through the broad-based support structure that has developed for the field.
- The panel recommends that the responsiveness and value of the field be further strengthened by developing closer ties with those areas and agencies that benefit and stand to further benefit from AMO science but that have not traditionally had strong links with the field, such as health, transportation, and environment. Institutions and agencies concerned with progress in these areas should also participate in the funding of AMO science.
- The panel strongly recommends that support for basic research be maintained at least at the current levels. The ability of the field to make innovative

---

[3] Control and manipulation apply to internal states as well as energy, time, momentum, and position and include the invention and development of methods to enable measurements of atoms, molecules, charged particles, and light at higher resolution.

[4] New sources of light include new types of lasers and related optical technologies but are not limited to these systems. The emphasis here is on new ways to produce light with well-defined characteristics—wavelength, power, pulse width, coherence, and other properties.

advances in strategic areas depends strongly on maintaining a healthy level of support for basic research. The National Science Foundation (NSF) has recognized the support for basic research, including that in AMO science, as one of its primary responsibilities. DOD, NIST, and DOE also have maintained reasonable support levels for basic research in AMO science to ensure a flow of ideas and talent into their strategic missions.

- The panel recommends that specific priorities that guide the support of particular areas of AMO science continue to be based on intrinsic scientific and technical merit as well as on the strategic mission objectives of the agency, industry, or other organization funding the activity. Increased support of a particular area should reflect unusual promise to advance the science and/or its potential application to national needs.
- Because of the diversity of AMO science and of the areas that benefit, it is impossible to make a single, detailed, linearized list of priorities. The panel recommends that a series of workshops be held, perhaps under the auspices of the National Research Council (NRC), each focusing on the role of AMO science in addressing priorities associated with particular national goals, clarifying and making known the mission objectives to the community, and identifying the most important ways in which the community can respond. Such workshops could be held, for example, in industrial and manufacturing technology, high-performance computing and communications, energy, the environment, and health.
- The panel recommends that all the federal agencies that fund AMO science actively support and participate in an interagency advisory or coordinating committee to collect and disseminate information about AMO science and its role in federal strategic initiatives and other areas of application and to provide guidance to government, industry, academic institutions, and others in the AMO science community. This committee should work closely with program managers of the federal agencies and should include representatives from a variety of constituencies, including AMO scientists and the end users of the research in industry and government. Such a communications link is particularly important in a diverse field with multiple sources of support and a broad range of end users. This need could be satisfied by expanding the breadth of the membership and activities of the NRC Committee on Atomic, Molecular, and Optical Sciences (CAMOS). The workshops mentioned above could be a major tool of the committee.

GENERAL RECOMMENDATIONS

- The panel recommends that the federal agencies emphasize support for single investigators and small groups and rely on merit review for exploratory as well as strategic, goal-oriented basic research.
- The panel recommends that academic institutions consider making changes

in the curricula, degree offerings, and advice they offer to students to make them more aware of and better able to respond to career opportunities and requirements in the many areas that are enabled by AMO science. Achieving this goal could be accomplished in part through greater interaction and cooperative programs with industry and government. Such changes should be designed to promote the interest of women and minorities in AMO science and increase their representation in the field. These issues, of course, transcend the particular field of AMO science and should properly be addressed more broadly.
- Given the narrow definition of optical science adopted in this study, the panel recommends that a more comprehensive assessment of the more broadly defined field of optical science, engineering, and technology be undertaken.[5]

---

[5] The National Research Council has approved the conduct of a major study of the field of optical science and engineering.

Part II

Atomic, Molecular, and Optical Science: Today and Tomorrow

The chapters that follow expand on the brief discussion of the field contained in Part I. Chapter 1's three "case studies" provide selected examples of how the results of basic atomic, molecular, and optical (AMO) research have provided, or promise to provide, major economic and human benefits. Chapter 2 highlights a number of recent advances in AMO science, together with its impact on other fields of science and engineering. This chapter also includes a discussion of the importance of AMO science to the nation's economy, security, and infrastructure. The topics of education and human resources, and of funding and institutional and organizational infrastructure for research and development in AMO science, are addressed in Chapters 3 and 4, respectively. The economic impact of the field is considered in Chapter 5. Chapter 6 touches on AMO science in the United States compared with that in other countries. Appendix A enumerates the Nobel Prizes awarded in AMO science since 1964 to further illustrate some of the accomplishments of the field. Appendix B presents a graphical representation of the broad impact of AMO science. Appendix C provides a citation analysis as one means of illustrating the U.S. position in this field. Results of the survey sent to nearly 20,000 scientists, as well as the survey itself, are discussed and presented in Appendix D.

The examples in this part of the report illustrate the health and vitality of AMO science, its interfaces with many other fields of science and engineering, the inherently interdisciplinary nature of much of the research, and the diversity of the AMO scientific community. The various discussions also highlight the importance of the discoveries, data, and technologies provided by AMO science in the many areas of commerce, industry, and human welfare so important to our nation.

1
Case Studies in AMO Science

In this chapter the panel presents three case studies that illustrate the close coupling between basic research, development, and application typical of AMO science. New ideas and discoveries made in the research laboratory are incorporated, sometimes rapidly, into new technologies and products and frequently enable advances in other areas of science. This diffusion of ideas maximizes the return on the initial research investment and is such that today the products of AMO science are an important contributor to the national economy.

The case studies also illustrate the different phases in the evolution from basic research to marketable products. The case study on lasers shows how basic research and development in AMO science has resulted in products with numerous commercial applications. That on optical manipulation of atoms highlights an area that has come into being only in recent years but in which commercial products are already beginning to appear. The case study on carbon nanostructures describes an exciting new area in which potential economic impacts are in the speculative stage.

LASERS: FROM BASIC RESEARCH TO NEW TECHNOLOGIES AND NEW INDUSTRIES

Lasers, one of the most remarkable products of twentieth-century science and technology, evolved directly from basic AMO research concerning light and its interaction with matter, including atoms, molecules, and solids. The laser has revolutionized many fields of science and is a device whose applications today touch all our lives. For example, long distance telephone calls are now routinely carried by laser beams traveling through optical fibers thinner than a human hair. Lasers have become indispensable weapons in the arsenal of medical therapeutic and diagnostic procedures, frequently providing attractive lower-cost alternatives to conventional surgery. Low-cost lasers have been developed, allowing their application in a wide variety of consumer products, including compact-disk (CD) players and laser printers. The total annual revenue of U.S. laser manufacturers is now approximately $1.1B (D. Kales, "Review and Forecast of Laser Markets: 1993," *Laser Focus World* **29** (January), 70-88, 1993), but the true economic impact of the laser is at least 100 times greater because lasers are a critical component of many consumer products and services.

The widespread application of lasers is a result of the unique characteristics of the radiation they provide. The output beam from a laser can be ultradirectional. Thus small objects can be selectively illuminated, even at large distances. This is key to the development of laser rangefinders, now widely used in surveying and in military applications, and of laser target designators. The pinpoint accuracy of "smart" weapons guided to their targets by scattered laser light was graphically demonstrated during Desert Storm. Laser beams may also be focused to extremely small spots, resulting in high energy densities sufficient to melt or vaporize many materials. This capability has resulted in numerous applications in, for example, industrial processing. Today lasers are used to heat, cut, or weld a wide variety of materials, including metals, ceramics, plastics, wood, and cloth. Radiation from a laser also can be highly monochromatic; that is, it encompasses a narrow range of frequencies and wavelengths. The output wavelengths available and their range of tunability vary with laser type. This wavelength choice makes possible spectroscopic analysis of atomic and molecular species and is the basis of many techniques used in remote sensing of atmospheric pollutants and of species present in environments such as combustion chambers.

Since the initial invention of the laser, research and development activities have resulted in the discovery of many different classes of lasers that, taken together, provide an enormous range of output wavelengths, pulse lengths, and power levels. A variety of lasers based on electrical discharges in gases have been developed, a good example of which is the helium-neon laser that provides the red beam observed in the point-of-sale scanners at the checkout stands in supermarkets and other stores (Figure 1.1). Argon and krypton ion lasers are widely used in ophthalmology and laser light shows, and carbon dioxide lasers find application in laser surgery. Many laser systems employ solid-state gain media including the ruby laser, which was the first system to demonstrate successful lasing. Recently, solid-state laser systems have been developed at infrared wavelengths that are "eye-safe," that is, that are absorbed by the fluid in the eye without damage to the retina, and these are finding applications in, for example, rangefinders and remote sensing. Laser systems based on organic dyes also provide tunable radiation but at wavelengths extending down into the ultraviolet.

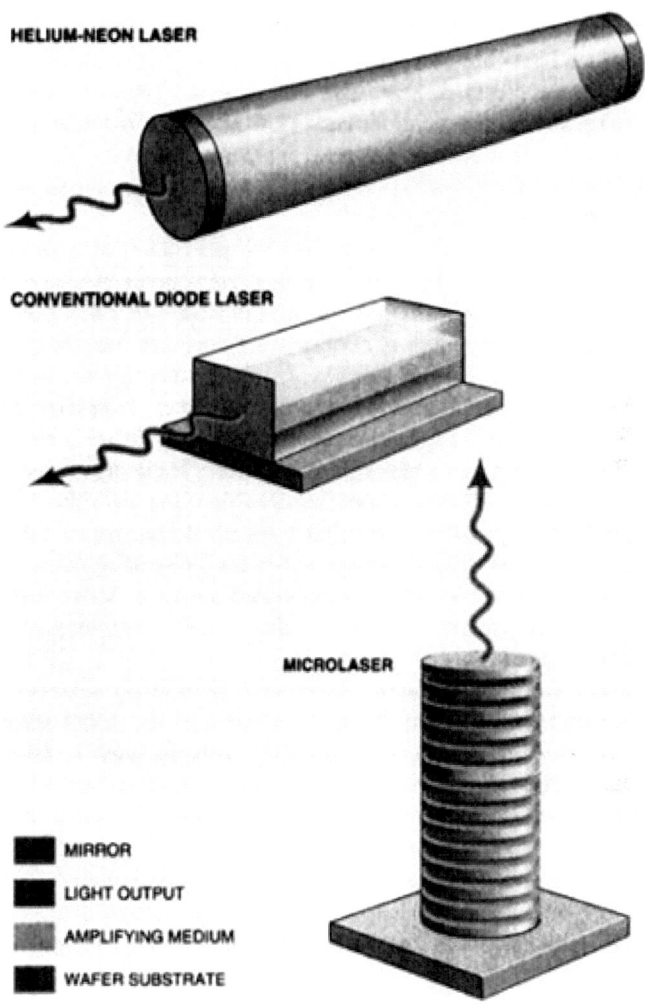

FIGURE 1.1 Lasers have a staggering diversity of uses and specifications. Shown here are a conventional helium-neon laser found in a supermarket bar code scanner (top), a conventional semiconductor diode laser used in a compact-disk player (middle), and a recently developed microlaser (bottom). These lasers range in size from hand-held (top) to about one-fiftieth the diameter of a human hair (bottom). Some lasers used for energy and defense research fill an entire large building. Just as the sizes vary (by over a factor of 10 million), so also do the functions and uses, encompassing commercial applications, manufacturing, medicine, transportation, environment, communications, defense, and scientific research. (Reprinted, by permission, from Jack L. Jewell, James P. Harbison, and Axel Scherer, "Microlasers," *Sci. Am.* **265** (November), 86-94, 1991. Copyright © 1991 by Scientific American, Incorporated. All rights reserved.)

Dye lasers have enabled major advances in spectroscopy and metrology. Semiconductor diode lasers are now available that are compact (typically the size of a grain of sand), simple to operate, and inexpensive. These are employed, for example, in optical communications and in CD players. Chemical, gas dynamic, and free-electron lasers provide powerful sources of infrared radiation with military applications. X-ray laser systems are also under development that promise advances in imaging and lithography.

Much effort has been directed toward the generation of high powers and very short pulse lengths. The search for alternate sources of energy has led to the development of lasers with pulse energies in excess of 1 kilojoule as part of the effort to harness the energy of nuclear fusion through ignition of thermonuclear fuel by laser implosion of small pellets. Such high-power lasers are also used as drivers for X-ray lasers and to produce extremely high temperature plasmas. Research has resulted in lasers that produce output pulses with durations as short as a few millionths of a billionth of a second. These enable measurements of processes occurring on time scales inaccessible using any other approach. It is, for example, now possible to monitor a chemical reaction as it occurs. The generation of ultrashort-duration pulses also holds the promise of extremely high rate communications and information processing systems. Major improvements in laser output beam quality, efficiency, stability, and reliability have been realized in recent years.

The many advances in laser systems and technology achieved in recent years have resulted from close interactions between the AMO sciences, condensed matter physics, electrical engineering, materials science, and traditional optics. Much research has been directed toward development of lasers with operational properties optimized for specific applications. These development projects, however, have frequently resulted in spin-offs into other areas. For example, eye-safe near-infrared solid-state lasers were initially developed for Department of Defense (DOD) and National Aeronautics and Space Administration (NASA) applications but now play an important role in medicine because their radiation is strongly absorbed by tissue.

The benefits provided by laser technology will continue to increase as new applications are discovered and as new highly reliable, efficient, easy-to-use, and moderately priced devices become available. In this regard, semiconductor diode lasers and solid-state lasers pumped by semiconductor lasers appear particularly attractive because of their modest size and cost and because of the ruggedness of solid-state technology for applications outside a laboratory. The last decade has witnessed the discovery and practical realization of solid-state lasers containing transition metal ions, including the titanium:sapphire laser, which is continuously tunable over a broad range in the near infrared and has become the laser of choice for the generation of ultrashort pulses. This development has been made possible in part by the high-quality crystals now available in the United States. The wavelength domain of semiconductor lasers continues to expand with the recent demonstration of lasers in the blue/green region. Indium gallium arsenide (InGaAs) and indium gallium arsenic antimonide (InGaAsSb) lasers have also been fabricated that extend wavelength coverage farther into the infrared.

With further research and development, solid-state lasers should evolve to the point that they will be efficient sources of laser radiation over the entire optical spectrum from the ultraviolet to the middle infrared. Their becoming efficient sources over this regime will require a combination of new materials, new devices, and new optical configurations and the use of nonlinear optics for frequency conversion. Semiconductor diode laser technology is also rapidly advancing on several fronts. Diode laser arrays are now being developed that provide high average powers and that can be used either alone or as pump sources for solid-state lasers. One interesting recent development is the creation of large arrays of laser diodes using techniques employed in very-large-scale integrated (VLSI) circuit technology. It may be possible to coherently lock the arrays, and if this is accomplished, diode lasers could eventually replace optically pumped solid-state lasers for many applications. Advances in semiconductor materials technology are making possible layered semiconductor systems (quantum wells) whose properties might be specifically tailored to optimize laser operation. Of particular interest are arrays of surface-emitting vertical cavity diode lasers. Unlike conventional, edge-emitting diode lasers, the surface-emitting lasers can be fabricated close together on a substrate. In the past few years, it has become possible to make arrays of about 1 million surface-emitting quantum-well lasers on a single 1-cm^2 GaAs substrate, each individual laser being only a few micrometers across and a few micrometers high. One of the long-term goals of semiconductor laser technology is the full integration of lasers and detectors with electronic circuitry, all on the same chip, to form optoelectronic integrated circuits.

Applications of optical fibers rely on their ability to guide a beam of light. A serious practical consideration is the attenuation of the light beam as it propagates along the fiber, an effect that becomes increasingly important as the fiber length increases. Recently, it has become practical to dope optical fibers and to pump them by injection of light from a diode laser in such a way that they amplify light by stimulated emission. Thus laser action has been realized in fibers. The erbium-doped silica fiber laser offers high gain together with the advantages of single-mode guided-wave fiber optics, making it extremely valuable as an amplifier or repeater for fiber-optic communications systems.

From their beginnings as a scientific breakthrough, lasers have become a prominent component of a large commercial market. They provide vital enabling technology in many areas critical to the nation's economic productivity, competitive position, and security. The applications noted above, and described in other chapters of this report, speak to their many contributions in areas such as manufacturing, materials, environmental monitoring, medicine, and communications.

It is difficult to assess the economic impact of the laser. Although the laser manufacturing business is relatively small, the economic activity that results from direct application of lasers in consumer products and in equipment that generates other goods and services is much greater. For example, sales of laser printers amount to $6.3B per year, and sales of CDs to $3.9B per year. The telecommunications market, in which the laser is a key player (together with other optical technologies), is now $161B per year. It is thus clear that the laser has had a substantial economic impact, and it is certain that as new laser systems and applications are developed this impact will continue to increase.

The story of the laser illustrates how investment in basic and applied research can lead to new technologies of great benefit to the nation and its people.

MANIPULATING ATOMS: NEW TECHNOLOGY FOR TODAY AND TOMORROW

Traditionally, neutral objects are held and moved by mechanical means. In the last few years, however, it has become possible to manipulate atoms and neutral microscopic particles with astonishing facility using light. What distinguishes this approach from other manipulation techniques is that it permits neutral particles to be positioned and moved without physical contact. The story of optical manipulation illustrates the interplay between basic research and the development of techniques with potential for widespread application in many areas, including ultraprecise atomic clocks, structural engineering at the atomic level (nanotechnology), and the study of deoxyribonucleic acid (DNA). The basic research was driven initially in large part by the desire to eliminate uncertainties caused by atomic motion and thus allow precise measurements on atomic systems. The goal was to be able to hold or trap an atom at rest in free space, unperturbed by contact with other atoms. The techniques developed to accomplish this grew out of a basic understanding of optical interactions with atoms that had been gained through decades of fundamental research.

Laser Trapping and Cooling of Atoms and Ions: Particle Optics

In 1978, it was suggested that atoms could be trapped at the focus of a laser beam. The trapping forces, however, were predicted to be very feeble and insufficient to overcome the normal random thermal motion of an atom. In order for trapping to be possible, the atoms would have to be cooled to a temperature of only a few thousandths of a kelvin, that is, only slightly above absolute zero. A technique to cool atoms with light was suggested in the seventies and had been demonstrated with charged atomic ions, but it was not until 1985 that neutral atoms were first successfully cooled to less than one-thousandth of a kelvin with a configuration of laser beams termed "optical molasses." Because of the Doppler shift resulting from the atom's motion, the light in this configuration acts as a viscous medium and atoms imbedded in the "molasses" are slowed to very low velocities. Once such cooling was obtained, trapping became fairly straightforward, and a variety of optical, magnetic, and magneto-optic traps were quickly demonstrated. Currently, scientists are able to cool trapped atoms to an effective temperature of less than 10^{-5} kelvin.

Concurrent with the development of laser cooling and trapping, neutral particle optics equivalent to lenses, mirrors, diffraction gratings, and beam splitters have been devised. Atom optics are especially powerful when coupled with laser cooling techniques because methods that would normally be rejected as too feeble to manipulate atoms at room temperature become feasible with slowly moving atoms. It is now possible, for example, to focus an atom beam by reflection from a curved surface (mirror) or to diffract and deflect a beam using a standing lightwave. The application of atom optics in metrology and lithography is already being explored.

The ability to manipulate cold atoms has also enabled scientists to develop devices that have no optical counterpart. For example, an "atomic fountain" in which laser-cooled and trapped atoms were tossed upward and shown to return in a ballistic trajectory due to gravity has been demonstrated. Unperturbed atoms in free fall make possible measurement times that are much longer than can be achieved by using conventional thermal atomic beams, and this achievement has stimulated work to build an improved atomic clock based on an atomic fountain. Work with laser-cooled ions, or even a single ion, in an ion trap also shows great promise for improved atomic clocks. The expectation is that the time standard used by the world today can be improved at least 100-fold. This refinement will affect many areas of science and technology. For example, the high-speed computers used in telecommunications are synchronized by atomic clocks. The most accurate means of global positioning, the Global Positioning System, depends on a set of satellite-based atomic clocks, and the highest-resolution radio images of distant galaxies yet obtained depend on the synchronization of signals received simultaneously by several radio telescopes.

Another device that has been made possible by a combination of atom manipulation techniques is the atom interferometer, which takes advantage of the wavelike nature of atomic particles. Conventional optical interferometers are used as extremely sensitive measuring devices, and interferometers based on atoms promise new measurement capabilities and increased precision. The first successful atom interferometers were reported only recently. One of these employed laser-cooled atoms in an atomic fountain. The long measurement time afforded by the fountain enabled measurement of the local acceleration due to gravity to high precision. Indeed, with further refinement, the interferometer may be comparable to the best absolute gravity meter available. Applications of this instrument include oil exploration, the measurement of land height changes due to the motions of continental plates that cause earthquakes, and monitoring of sea height changes due in part to global warming and the melting of the polar ice caps. A slight modification of the geometry of the gravity meter may also allow construction of an extremely sensitive atom gyroscope that could greatly increase the accuracy of inertial navigation systems.

Optical Tweezers and the Biosciences

The work on atom trapping stimulated renewed interest in the manipulation of microscopic neutral particles. Several years prior to demonstration of optical trapping, micron-sized glass spheres were levitated and trapped in air by laser beams, but it was not until much later that it was realized that a similar beam could be used to trap the spheres in water. The trapping forces are strong enough to overcome thermal motion at room temperature. The great advantage of using a single beam is that it can be used as an "optical tweezers" to manipulate small particles. The optical tweezers can be easily integrated into a conventional microscope by introducing laser light into the body of the microscope and using the microscope objective to focus the light beam (Figure 1.2). A sample can be viewed and simultaneously manipulated by simply moving the focused light beam.

Shortly after the demonstration of the light trap, it was discovered that bacteria could be captured in the trap, moved at will, and then released. If the trapping light is in the near infrared (where compact and inexpensive lasers are available), the bacteria can be held without apparent optical damage. Since that discovery, scientists have moved rapidly to apply this tool to a variety of problems. Scientists have reached inside living cells and moved organelles in the cytoplasm (Figure 1.3), studied the mechanical properties of the molecular motor that propels *Escherichia coli* through water, and used a laser beam to cut chromosomes within the nucleus of a cell undergoing mitosis, the fragments subsequently being moved into different locations using the optical tweezers.

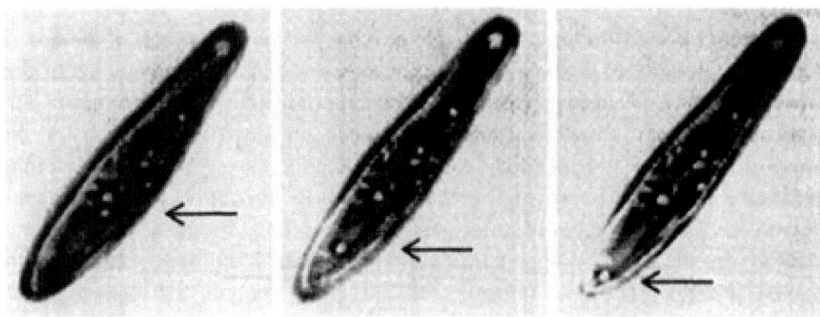

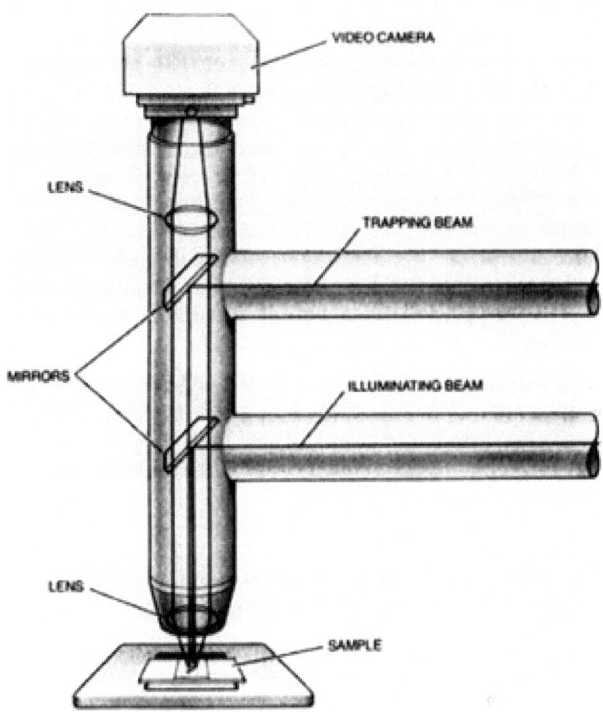

FIGURE 1.2 Optical tweezers can manipulate microscopic objects such as cells. A sample is placed on the stage of a microscope, which has been adapted to admit green laser light and infrared laser radiation. The green light illuminates the sample while the infrared radiation raps and holds it. (Reprinted, by permission, from Steven Chu, "Laser Trapping of Neutral Particles," *Sci. Am.* **266** (February), 71-76, 1992. Copyright © 1992 by Scientific American, Incorporated. All rights reserved.)

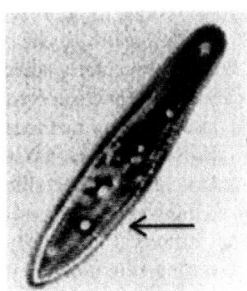

FIGURE 1.3 Organelle inside a protozoan was dragged to one end of the cell using an optical tweezers, as shown in the first three photographs. The image seen at the far right shows the organelle after it was released. (Photographs courtesy of Arthur Ashkin, AT&T Bell Laboratories. Reprinted from Steven Chu, "Laser Trapping of Neutral Particles," *Sci. Am.* **266** (February), 71-76, 1992.)

Only a few years after laboratory demonstration, commercial versions of optical tweezers are already available.

On a still finer scale, individual molecules of DNA can be manipulated by two optical tweezers. Although the molecule is too small to be held by an optical trap at room temperature, it is possible to attach micron-sized polystyrene spheres to the ends of the molecule, which act as tiny handles that can be held by the optical tweezers. The elastic properties of the molecule have been measured with this technique, testing basic theories of polymer physics. Work is under way to stretch out and then pin down the molecule so it can be examined by an atomic force microscope. The ability to manipulate the molecule while viewing it in an optical microscope (DNA can be stained with a fluorescent dye) opens up a number of exciting possibilities to study the function of gene expression, regulation, and repair.

Optical manipulation is an example of how basic research on the interaction of radiation with matter and the trapping and cooling of atoms has resulted in powerful new tools for physics, chemistry, geology, and the biosciences with many applications.

BUCKYBALLS AND CARBON NANOTECHNOLOGY: SURPRISING NEW MATERIALS FROM SMALL SCIENCE

Since the 1800s, students have learned that elemental carbon naturally occurs in two forms: graphite and diamond. In graphite, carbon atoms are bonded together in large sheets, and in diamond carbon is bonded in a three-dimensional tetrahedral crystal structure. However, recent and unexpected discoveries have shown how to make carbon in a number of other structural forms with a wide variety of different and potentially useful mechanical and electrical properties. AMO research has revealed a broad new class of materials with major importance in materials science, solid-state physics, and chemistry.

The remarkable discoveries described here grew out of research programs designed to understand the physical properties of microscopic nanometer-sized particles and clusters. A cluster is a molecule or particle consisting of approximately 10 to several hundred atoms. Theory had predicted that many physical properties would be sensitive to the size of a cluster, but experimental studies became possible only with the invention of new techniques. To form clusters of refractory elements normally found only as bulk solids, laser ablation methods were developed, and laser spectroscopic methods were devised to characterize their structure. A major fraction of this effort was devoted to clusters of metallic and semiconducting compounds because it was recognized that the strong bonding of these materials might cause the clusters to adopt different, previously unseen, structures and types of bonding. For example, it was quickly found that clusters of alkali metals, where bonding is essentially isotropic (that is, nondirectional), could be described by a simple electron shell model partially analogous to the nuclear shell model. As a result, clusters containing certain numbers of electrons have unusual stability and abundance.

Carbon and, to a lesser extent, silicon represent the opposite chemical extreme from alkali metals in that the bonding is strongly *directional* in space. Simple theoretical considerations and early free radical experiments suggested that small carbon clusters would exist as chains and rings, but there were no reliable predictions of how large a carbon cluster would have to be to exhibit graphite or diamond structure, or what intermediate-sized clusters might actually look like. Carbon, then, was expected to be an intriguing cluster research problem, and in 1985 it was observed quite unexpectedly that a carbon cluster with 60 atoms is far more stable against growth to larger species than are clusters of neighboring numbers of atoms. It was proposed, without direct structural evidence or a macroscopic sample of the material, that this C_{60} cluster had the now famous truncated icosahedral structure. This icosahedral structure, and its popular name "buckyball," was inspired by the geodesic domes of R. Buckminster Fuller. The connected structure of the figure implies that C_{60} is a stable molecule with strong bonding on the surface of the icosahedron. Conceivably, other molecules and atoms could be chemically bound to the surface, or could even be trapped *inside* the icosahedron.

This proposal was revolutionary, yet, as occurs in most significant discoveries, it is simple and elegant in retrospect. It immediately suggested that C_{60} would be a sufficiently stable molecule to be the building block of new materials, essentially a new form of elemental carbon. The probable existence of C_{60} became widely known in chemistry, materials science, and solid-state physics, and for 5 years scientists tried without success to find a practical method to make macroscopic amounts of C_{60}.

A practical method came from a totally unexpected source. Researchers studying light scattering from graphitic particles recognized that C_{60} might account for the curious optical spectra associated with "soot" made for decades by graphite rod vaporization in helium. C_{60} was crystallized using chemical methods from such soot and proved to have an icosahedral structure. As C_{60} is present in an astonishing ~ 25% yield, a practical synthesis was achieved.

Synthesis of C_{60} triggered a rapid series of further discoveries that continues today. Materials chemists studying organic conductors recognized that crystalline C_{60}, when doped with extra electrons, might be an organic metal with nearly isotropic conductivity. Organic metals are currently being developed for a number of materials applications in electronics and optics. To their great surprise, they found that alkali-doped C_{60} crystals are superconductors at low temperature. Theory has not yet advanced to the point that new superconductors can be predicted, and so the discovery of a new type of superconductor, along with the fundamental understanding of the microscopic mechanisms responsible for the superconducting, is important. Superconducting materials have important applications in integrated circuits, magnetic sensors, large magnets for imaging systems, and power transmission systems.

Buckyballs can now also be made with metal atoms inside the cage. The interior atoms donate their valence electrons to the cage and thus change its electrical properties. Organic molecules can be bonded to the outside of the cage, and this fact allows buckyballs to be incorporated into polymeric "plastics." The possibilities of both outside and inside doping of buckyballs in solids should give electrical properties we cannot now anticipate. The C_{60} structure, however, is now known to be the smallest member of a family of similar, larger icosahedral structures, the next member of which is C_{240}. Remarkably, these structures can occur nested in an "onion skin" arrangement, and their properties are only just beginning to be examined.

In the past year another dimension of carbon nanotechnology was revealed

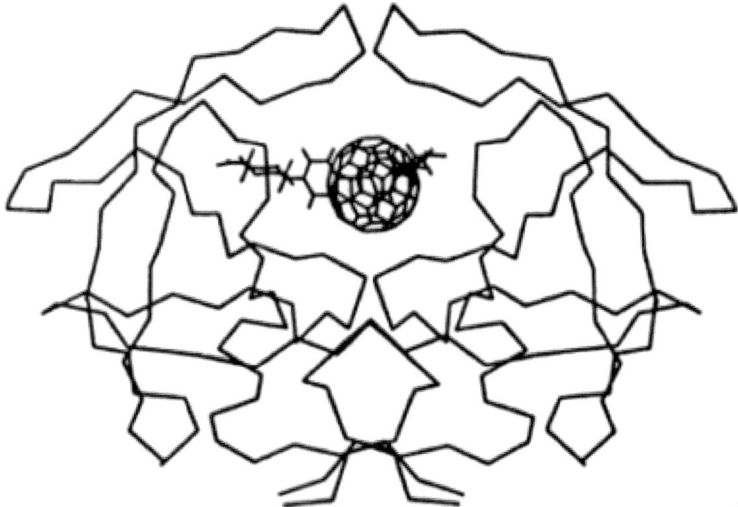

FIGURE 1.4 Fullerenes are expected to present many opportunities for new materials, including, perhaps, some of the strongest fibers known when "nanotube" technology is fully understood and exploited. Here another possible use is illustrated; specifically, a buckyball is shown attached to the AIDS (HIV) virus, neutralizing a very large number of the active sites of the virus. (Reprinted, by permission, from S.H. Friedman, D.L. DeCamp, R.P. Sijbesma, G. Srdanov, F. Wudl, and G.L. Kenyon, "Inhibition of the HIV-1 Protease by Fullerene Derivatives: Model Building Studies and Experimental Verification," *J. Am. Chem. Soc.* **115**, 6506-6509, 1993. Copyright © 1993 by the American Chemical Society.)

when cigar-shaped, capped tubes of 1- to 2-nanometer diameter, and even extended "nanotubes" of essentially macroscopic lengths, were experimentally discovered. Nanotubes are, in essence, sheets of graphite that have been rolled into tiny tubes. It is predicted that they can be either semiconductors or metals, depending on the angle of the bonding with respect to the tube axis. Nested nanotubes have also been recently observed.

More broadly, the discovery of buckyballs and the carbon variants has stimulated researchers in many other fields to consider a wide range of useful nanostructures that might be built from curved graphite sheets. Researchers are now actively pursuing the growth of continuous carbon nanotubes. When grown in length to many meters, these nanotubes would constitute a fiber of incredible tensile strength. Multifilament cables of such pure carbon nanofibers are expected to be the strongest (and toughest) cables that could ever be constructed of any material—roughly 100 times stronger than a steel cable of the same diameter and 400 times stronger than steel per unit weight. Nanofibers, doped with metal atoms down the hollow inside cavity, are expected to have electrical conductivity at room temperature substantially higher than pure copper and could provide an attractive replacement for use in electrical power transmission lines worldwide. Furthermore, their large thermal conductivity suggests they may replace diamond and copper in many applications requiring heat sinks. AMO science was critical to the genesis of these ideas and will be critical in bringing them to life.

The importance of C_{60} is extending into research areas beyond those associated with new materials. For example, in biomedical research, recent theoretical and experimental tests have shown that C_{60} derivatives interact with the active site of HIV-1, removing much of the active area of this deadly virus (Figure 1.4). Thus it is conceivable that this recently discovered particle may offer some help in the worldwide battle against AIDS.

> *Carbon, already important in biology and organic chemistry, now offers the possibility of new materials, devices, and products with numerous potential applications.*

2

Recent Major Advances and Opportunities in AMO Science and Applications to the Needs of Society

The intellectual excitement and rate of progress in the field of AMO science are at an all-time high. The discovery and development of the laser and of other innovative techniques make this a time of unparalleled scientific opportunity, and these discoveries have led to scientific and technological advances that once could only be dreamed of. AMO science has enabled major technological advances in manufacturing, materials, communications, space, defense, energy, the environment, health, and transportation that have had a major impact on the nation's economic productivity, competitive position, security, and technological infrastructure and on the general well-being of its people. This chapter provides examples that highlight recent advances in AMO science and its applications and that illustrate the promise of the field.

Some of the examples included below have been discussed in a recent National Research Council (NRC) report, *Research Briefing on Selected Opportunities in Atomic, Molecular, and Optical Sciences* (National Academy Press, Washington, D.C., 1991). More detailed discussions of many of these topics can be found in the report *Future Research Opportunities in Atomic, Molecular, and Optical Physics* (PUB-5305, Lawrence Berkeley Laboratory (LBL), Berkeley, California, 1991), sponsored by the Department of Energy (DOE). Here, as in the earlier reports, it is only possible to touch on a few of the accomplishments and opportunities because their number is so large.

THE NATION'S SCIENTIFIC KNOWLEDGE BASE

New and revolutionary discoveries of a fundamental nature continue to be made in AMO science, and, as demonstrated by the citation analysis presented in

Appendix C, the United States is a world leader in this area. To identify scientific forefronts, technological opportunities, and windows of future opportunity, broad input was sought from the AMO science community through letters to individual scientists, open sessions at professional society meetings, and other forums. That input identified the areas presented below as those meriting special discussion. Clearly, not all areas of significant progress can be included in such a brief presentation. Nevertheless, this summary will provide a glimpse of some of the scientific frontiers in AMO science.

Recent Discoveries and Future Opportunities in AMO Science

Fundamental Laws and Symmetries

One unique aspect of AMO physics, distinguishing it in all of science, is the capacity to make measurements with extraordinarily high precision. In suitably chosen systems, such precision measurements can probe physics far beyond the confines of what is customarily considered AMO science. For example, precision AMO measurements are testing our basic concepts of space and time, revealing new details about nuclear structure, probing the existence and properties of elementary particles, and exploring our fundamental understanding of the forces of nature. In this way, AMO science provides the unusual opportunity to explore the frontiers of physics without leaving the proverbial "tabletop." A common theme in much of this work is that one tests accepted theories at increasingly higher levels of precision until, at some point, a discrepancy is observed, which leads to important new insights.

One example is the testing of the ideas of space and time that are embodied in the theory of special relativity. The optical experiments of Michelson and Morley and of Kennedy and Thorndike provided important early tests of the isotropy of space and the speed of light. Recently, laser versions of these experiments have tested the isotropy hypothesis at a precision many orders of magnitude higher. Similarly, recent laser spectroscopy experiments have provided dramatically improved precision for the confirmation of the time dilation formula of special relativity.

The area of precision measurements that has made perhaps the largest contribution to the basic understanding of physics is the detailed examination of atomic structure. Historically, precision measurements of atomic and molecular spectra laid much of the groundwork for the development of quantum mechanics, and the high precision of these data provided an exceptionally rigorous testing ground for new theories as they were put forth. The microwave technology developed in World War II led to the precise measurement of the Lamb shift in hydrogen, which stimulated the modern development of quantum field theory. The rather radical concepts of renormalization and vacuum fluctuations gained quick acceptance because of the remarkable agreement between theoretical predictions and experiment. The development of quantum electrodynamics has advanced to the point that calculated and measured values of the magnetic moment of the electron agree to 1 part in 10^9.

This same spirit underlies the ongoing efforts to find and study new, non-Coulomb forces through their effects on atomic structure. Particular attention has been given to forces that violate time (T) and/or parity (mirror) (P) reversal symmetries. At present the only evidence for a time-reversal-violating force is in the neutral kaon system, and there is much speculation, but no solid arguments, as to its origin and relationship with other forces. An electric dipole moment (EDM) of a fundamental particle could only exist if there were a T-violating force, so there has been a long history of experiments that have searched for EDMs of neutrons, electrons, atoms, and nuclei. Over the years, these measurements have improved enormously and now have achieved extraordinary levels of sensitivity. For example, the present limit on the EDM of an atom is equivalent to a displacement of the positive and negative charges of about 10^{-26} cm. The fact that EDM searches have yet to yield a nonzero result places severe constraints on theoretical models. Indeed, the standard model is one of the few that have been put forth that can simultaneously explain the presence of T-violation in kaons and its absence at the level currently set by experiments in atomic systems. It is generally believed, however, that the standard model is only a part of a larger scheme, such as supersymmetry. EDM searches are one powerful means to test many of the ideas put forth in these larger schemes.

The study of parity nonconserving forces in atoms is, in a historical sense, positioned between the Lamb shift and the search for the electric dipole moments. The Lamb shift was at one time a radical new discovery but is now accepted without question, whereas the search for EDMs is a quest (as yet unfulfilled) to observe a new phenomenon. Throughout this century, there have been numerous experiments that set limits on parity nonconservation (PNC) in atoms, but it has only been in the last 10 to 15 years that experimental precision reached the necessary level to detect this effect. Although atomic PNC is at the extremely small level of only 1 part in 10^{11} mixing of atomic states, it is now possible to measure the size of this mixing, for example in a cesium atom, to 2% accuracy. PNC in atoms is understood to be a manifestation of the neutral weak interaction that was predicted as a result of efforts to unify the weak and electromagnetic forces. The study of the neutral weak force remains one of the best methods of testing the standard model of electroweak unification and probing the many unexplained features of this model. By themselves, the experimental atomic PNC data are not sufficient to probe the nature of the neutral weak force. In order to interpret the data, it is also necessary to precisely calculate the structure of many-electron atoms such as cesium. In the past few years, dramatic advances in computational techniques have made this possible. This "marriage" of theory and measurement of atomic PNC now provides one the most precise tests of the standard model and complements the many high-energy tests of the model, because only the atomic experiments are sensitive to the values of two of the four basic electron-quark neutral current coupling constants. A substantial number of new theories have been put forth to avoid the problems of the standard model, and their primary observable effects involve these two coupling constants. Because of this, the atomic PNC results set the most stringent constraints on much of this so-called "new physics." As these experiments and the atomic structure calculations are improved, they will further probe the nature of the electroweak unification. Only a few other experiments have comparable potential to address basic questions in elementary particle physics, and these are generally being conducted within the high-energy physics community.

Precision AMO measurements of the sort discussed here drive, and are driven by, the technology of the entire AMO field. For example, the progress in the search for EDMs has directly reflected improvements in radio frequency resonance and, more recently, in laser spectroscopic techniques. Work is now under way to incorporate laser cooling techniques into these searches, which promises to lead to dramatic improvements. Efforts to improve special relativity tests and atomic PNC experiments have led to developments in laser stabilization and optics that are now spreading throughout the field of AMO science and, in particular, have had a substantial impact on techniques for laser cooling and trapping of atoms. This mutually beneficial relationship will continue to advance the "tabletop frontier," as well as further drive the technology that has made AMO science such a vital field.

Cavity Electrodynamics and Micromasers

Cavity quantum electrodynamics (QED) deals with the modification of free-space atomic radiation processes by cavities and other structures. Although it has been nearly 50 years since such effects were first considered, it is only in the past decade that experimental techniques, especially the use of Rydberg atoms and superconducting cavity walls, have become available to study such effects with single atoms. These techniques have allowed the observation of the exchange of energy between an atom and a single mode of the electromagnetic field in a cavity, which has been successfully modeled theoretically. When this treatment is extended to allow for the coupling of the atom to an arbitrary number of modes, the sinusoidal exchange of energy is generally replaced by an effectively irreversible transfer of energy from an excited atom to the field. The effect of the cavity is then to modify the atom's spontaneous emission in a way that depends on such things as the position of the atom within the cavity and the reflectivities of the cavity walls. Spontaneous emission can be inhibited if there are no allowed cavity modes at the emission frequency. Various experiments have verified the predicted modifications of radiative lifetimes by cavities.

Recent developments have also opened the possibility of experimental studies of the transition between small and large systems. Such studies promise to shed light on the few-body problem and classical/quantum correspondence. In so-called micromasers a low-density beam of Rydberg atoms is injected into a single-mode microwave cavity at such a low rate that at most one atom at a time is in the cavity. As such, micromasers are dynamically driven systems: both the cavity mode and the atoms are dynamical systems, so that the cavity mode in particular is an open system that can evolve from mixed states to pure states. Since good photon detectors in the microwave range are not available, one studies instead the state of the Rydberg atoms as they exit the cavity. The atoms play the dual role of pump and detector. This measurement scheme makes the micromaser a particularly attractive test system to investigate a number of aspects of quantum measurement theory. In addition to the issues in quantum dynamics and measurement theory, micromasers are theoretically attractive because the amplifying medium is relatively simple and an accurate quantum treatment is possible. Quantum fluctuations play an important role in these systems, since the mean photon number in the cavity is extremely low.

Laser action is possible in the micromaser because the field amplification by a single atom is sufficient to overcome the tiny loss of the cavity. It is possible using micro-optical, non-superconducting cavities to achieve laser action in many-atom systems with nearly arbitrarily low pumping levels, and without any distinct pumping threshold. This laser action occurs because in a micro-optical cavity, where one dimension is of the order of half a wavelength long, photons are emitted into a single mode, without spontaneous emission into all nonlasing modes, as in a conventional laser. Increased pumping then results in a gradual transition from predominantly spontaneous to predominantly stimulated emission, without a sharp threshold. Thresholdless lasing has been demonstrated with microcavities containing dye solutions and should be possible also with semiconductors. The ultralow power consumption of such devices makes them interesting for various applications. Microcavity lasers also offer response rates exceeding 100 gigabits (100 billion bits) per second, which cannot be realized with more conventional lasers.

Efficiencies of such important semiconductor devices as laser diodes and solar cells are strongly determined by electron-hole radiative recombination rates. Radiative decay rates can be suppressed in geometries having no allowed electromagnetic modes at the radiative wavelengths, as already noted. Inhibition can also be realized by forming structures in such a way as to produce photonic band gaps, that is, regions in a transmission versus wavelength curve where transmission is forbidden, analogous to forbidden energy bands of electrons in crystals. Recently, a cubic lattice with a photonic band gap having a width of about 20% of the central (optical) frequency has been fabricated. Such structures, which are an extension of the ideas of cavity QED, offer the possibility of dramatically improving the efficiencies of various electronic devices.

Although photonic band gap structures are currently in a basic research stage, it is not difficult to imagine possible practical applications. For instance, they might be used in lasers to inhibit radiative decay into nonlasing modes. This could be commercially important because it might lead to a substantial reduction in pumping requirements of diode lasers and contribute to the development of thresholdless lasers.

Highly Perturbed Atoms in Intense Laser and Microwave Fields

The interaction between laser light and atoms has been an active area of experimental and theoretical research since the discovery of the laser. By the early 1980s, these interactions were thought to be reasonably well understood, based on well-characterized perturbation theory. At about that time, however, improvements in laser technology led to intensities approaching 10^{13} watts per square centimeter (W cm^{-2}), where the ponderomotive, or "quiver," energy of the electron in the field becomes comparable to the photon energy. Intensities a few orders of magnitude higher are now possible, providing laser fields comparable to the strength of the electric fields that hold an atom together. At these intensities, where the effect of the laser light on the atom is not a small perturbation, entirely unexpected phenomena began to be observed. Experiments measuring the energy distribution of photoejected electrons showed large peaks corresponding to the absorption of many more photons than were necessary to ionize the atom. Such processes, now known as above-threshold ionization (ATI), had previously been thought to be of marginal importance but were found to dominate the spectrum at high laser intensities. Other unanticipated phenomena were also discovered, such as inexplicably high probabilities for multiple ionization of atoms by strong laser fields. Unexpected results had been observed earlier in studies of multiphoton ionization of atoms in high-lying Rydberg states by strong microwave fields.

It is now clear that these various experiments were signaling the entry of atomic physics into the realm of strongly coupled systems, where perturbation theories no longer can be depended on to provide descriptions of atomic behavior. Entry into this new realm pushed atomic theory in two different directions. The first was away from simpler perturbative approaches and into extensive computer analyses of the detailed quantum mechanics of these problems. Because these become extremely complex and difficult, a complete analysis is not yet possible even with present computational capabilities. However, calculations of appropriate model problems have now qualitatively reproduced most of the experimentally observed features. One intriguing prediction of these calculations is that with increasing field intensity in the strong field regime atoms can actually be stabilized against ionization. This counterintuitive prediction suggests that ionization probabilities are not, in all cases, monotonically increasing functions of field intensity.

The second theoretical approach to describing the behavior of atoms in intense laser fields has been through the broad field of nonlinear dynamics and chaos and, more specifically, through the investigation of quantum dynamics in systems with chaotic classical limits. These investigations, often referred to as "quantum chaos," mark a paradigm shift in atomic physics and include the study of statistics of energy-level spacings, the phenomenon of "dynamical localization" representing the suppression of diffusive behavior seen in the classical limit, and "scarring," the peaking of eigenstates of the time-evolution operator or quasi-energy states on unstable classical invariant structures. Because studies of nonlinear dynamics and chaos tend to emphasize universal aspects of the phenomena, this suggests that atomic physics might be an important testing ground for the development of new ideas having application throughout the physical world.

Microwave experiments involving Rydberg states of hydrogen have provided evidence for scarring and dynamical localization. These mechanisms provide the most complete and compact explanation of the experimental observations, while a more conventional interpretation is either exceedingly cumbersome or not possible. Recent efforts to extend the same approach to problems of ATI have led to the suggestion that scarring is relevant to the interaction of ground state atoms with intense lasers. The universality of the ideas of chaos and nonlinear dynamics links these studies to recent investigations of such seemingly unrelated areas of atomic physics as the spectrum of the diamagnetic hydrogen atom, the doubly excited spectrum of the helium atom, Rydberg charge transfer, and the motion of charged particles in traps. These well-characterized and controlled studies in atomic physics can then serve as paradigms for higher-dimensional problems in other areas, including atomic or molecular collisions that are not weak and cannot be considered slow or fast, driven quantum wells, and other mesoscopic solid-state systems.

Transient States of Atomic Systems and Collision Dynamics

The key to understanding a vast array of complex atomic collision phenomena involving the transfer of energy, angular momentum, and charge is, in many cases, the accurate description of the transient intermediate states of the collision complex. Improved experimental capabilities are making possible far more complete analyses of such states than have been heretofore possible, thereby permitting stringent tests of theory. In addition, a host of novel physical processes are being studied experimentally for which a theoretical understanding is only beginning to be developed. The scope of such processes encompasses all the various interactions between photons, electrons, positive and negative ions, neutral atoms, and even antiparticles.

Dynamics of Three-Body Systems. While systems that comprise two interacting particles may be described analytically, systems having three or more interacting particles can be described only approximately. Three-body systems are thus the prototypes of many-body systems. Experimental and theoretical progress in their description is a key to much of the physics of the everyday world. Indeed, the physics of many complex processes is governed by the interactions of three particles (one or more of which is often a composite particle). Important progress has recently been made in several areas.

High-resolution experimental measurements of the photodetachment cross section for the negative hydrogen ion have uncovered an extremely rich spectrum of doubly excited states, and this has been accompanied by commensurate theoretical advances. Experimental measurements of photo double ionization of atoms have revealed an intriguing empirical relationship between this process and electron impact ionization of the corresponding singly charged ions. The measurements are so precise that they have also tested theoretical understanding of the threshold laws for three-body breakup and, indeed, have led theory to predict alternative modes for three-body breakup applicable in different energy regions above threshold. Finally, the experimental measurements at high photon energies are providing severe tests of theoretical estimates of the ratio of double to single photoionization at asymptotically high photon energies.

Resonances in low-energy electron-atom and electron-molecule collisions provide detailed information on transient states of negative ions. Many such states are doubly excited, and electron-correlation effects are of paramount importance in their theoretical description. Several years ago, experimental data on electron-cesium scattering were used to make a semiempirical prediction of a stable ^3P state of the negative cesium ion. The existence of this state has more recently been predicted by ab initio theoretical calculations and confirmed by direct experimental observation.

Collisions at Ultralow Temperatures. Recent advances in laser cooling and manipulation of alkali, alkaline earth, and rare-gas metastable atoms have opened up many new opportunities in atomic collision studies. In particular, these advances now permit the study of inelastic energy transfer and associative and Penning ionization reactions at temperatures below 1 millikelvin (mK). The study of such collisions is important for both practical and fundamental reasons. Collisions limit achievable trap densities and give rise to difficulties in intended applications, such as preventing the realization of Bose-Einstein condensation of cold trapped atoms or causing pressure shifts in high-precision atomic clocks. Low-temperature collisions also display a number of unusual characteristics. Two distinctly different classes of collisions can occur. Collisions of ground or metastable states can be described by conventional scattering theory. In the near-threshold regime, only s-wave collisions have nonvanishing inelastic rates, and the rate coefficients are sensitive to the spin statistics of the colliding atoms and are subject to manipulation by external magnetic fields. In contrast, if laser radiation tuned near the cooling transition is present, collisions involving excited states will occur. These collisions are controlled by an extremely long range resonant dipolar interaction and enable a new kind of molecular spectroscopy, which probes the long-range potentials of the excited molecule formed from the two atoms and can be used to study molecular bound states having turning points of up to many hundreds of angstroms. Collision rates can be manipulated optically by varying the detuning and intensity of the radiation. If the detuning from resonance is only a few natural line widths, the atoms can be excited only when they are far apart, and they undergo optical pumping and spontaneous decay as they interact during their slow approach to one another. Conventional scattering theory no longer suffices, and new theoretical methods must be developed to account for the fluctuations and dissipation during the long-range part of the collision. Numerous experimental and theoretical studies of various aspects of the physics of ultracold collisions are now beginning, and this subject promises to have an exciting future.

Highly Charged Ions. Highly charged ions dominate hot plasmas such as those encountered in nuclear fusion reactors, X-ray laser research, and stars. Accurate spectroscopic and collision data are required to model the behavior of such plasmas. The technology for making and handling highly charged ions has advanced dramatically during the past few years, opening a whole new class of phenomena for study. Highly charged ions provide a critical test bed for our fundamental understanding of atomic structure and interactions. The enhanced long-range Coulomb forces between highly charged ions and other charged particles give rise to large cross sections for some processes. The electronic potential energy carried by the ion can overshadow kinetic effects in slow collisions. For highly charged ions, the reduced nuclear screening increases the binding energy of the outer valence electrons. Thus, processes involving inner-shell electrons often dominate as the ionic charge increases, for example, in electron-impact ionization. Electron-electron correlation effects may also be enhanced and elucidated in interactions of highly charged ions, for example, in multiple electron capture.

New Insights into Molecular Dynamics

Recent advances in laser technology have enabled scientists to examine phenomena with femtosecond time resolution. This capability has triggered many developments in AMO science. It is now possible, for example, to observe atoms moving in response to chemical forces, to monitor the flow of energy out of individual chemical bonds, and to contemplate optical control of the outcome of particular reactions (Figure 2.1). Complementary to this experimental understanding of molecular motion are numerous conceptual theoretical breakthroughs permitting, for example, the analysis of static spectra using time-dependent wave packets or the observation of flux propagation through curve crossing regions.

RECENT MAJOR ADVANCES AND OPPORTUNITIES IN AMO SCIENCE AND APPLICATIONS TO THE NEEDS OF SOCIETY

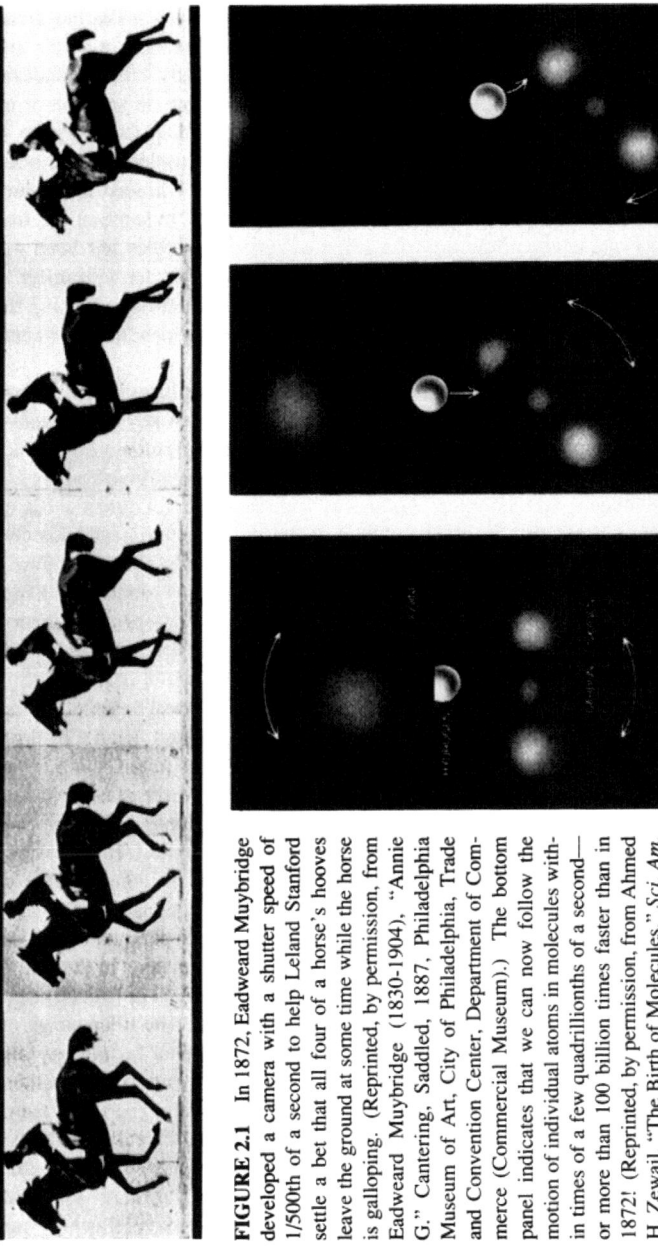

FIGURE 2.1 In 1872, Eadweard Muybridge developed a camera with a shutter speed of 1/500th of a second to help Leland Stanford settle a bet that all four of a horse's hooves leave the ground at some time while the horse is galloping. (Reprinted, by permission, from Eadweard Muybridge (1830-1904), "Annie G.," Cantering, Saddled, 1887. Philadelphia Museum of Art, City of Philadelphia, Trade and Convention Center, Department of Commerce (Commercial Museum).) The bottom panel indicates that we can now follow the motion of individual atoms in molecules within times of a few quadrillionths of a second—or more than 100 billion times faster than in 1872! (Reprinted, by permission, from Ahmed H. Zewail, "The Birth of Molecules," *Sci. Am.* **263** (December), 76-82, 1990. Copyright © 1990 by Scientific American, Incorporated. All rights reserved.)

The ability to excite and probe molecules on very fast time scales has greatly enhanced the understanding of energy transfer and relaxation in solids and liquids and of the dynamics of solvation. This is a particularly valuable endeavor because much of the important industrial chemistry occurs in solution or on catalysts. For example, these experiments have begun probing the key steps in solvation and reaction by monitoring the fate of an excited molecule in solution or as individual solvent molecules are added to clusters. Ultrafast techniques have also opened new possibilities for studying interfaces. In some of the first experiments, the energy flow out of surface-adsorbed molecules has been observed directly. These approaches provide new details about charge transfer in molecules adsorbed on electrode surfaces, and nonlinear spectroscopy using the high fields created by ultrashort pulse lasers permits entirely new measurements that reveal the nature of molecules adsorbed on surfaces.

Efforts to understand atomic and molecular interactions have produced fundamental new insights into reactive and nonreactive events. A key new direction involves the complete solution of the "vectorial" nature of collision dynamics problems, requiring elaborate angular momentum analyses and ingenious polarization and imaging technologies. Another is "coherent" control dynamics using preparation of nonstationary states. Many of the recent experimental insights come from the application of lasers to quantum state probing, the development of which has involved atomic, molecular, and optical sciences together. Computer technology has vastly expanded the scope of the accompanying theory, enabling the calculation of potential energy surfaces and dynamics through quantum and classical methods.

By applying lasers to a variety of elementary processes, researchers are exploring atomic and molecular interactions in unprecedented detail. In these experiments, atoms or molecules are prepared in selected initial states, even including selected orientations or alignments in space, and the states are analyzed after reaction or inelastic scattering. This preparation and analysis extends to the control of chemical reactions in a few prototypical cases. The measured orientation dependences provide the underlying basis for an understanding of reaction stereochemistry. One approach to controlling reactions is preparation of eigenstates that have special reactivity, and the other is preparation of nonstationary (coherent) states that allow laser control of the chemistry. In the future, it is envisioned that individual molecular bonds could be broken or formed selectively by the injection of modulated pulses of light with specific frequencies.

Rapid advances in theory, driven by both computational technology and new ideas, are being made. The implementation of time-dependent quantum mechanical methods for practical calculations is one example. The evolution of a quantal wave packet, properly analyzed, provides in one calculation the answers to numerous questions concerning the absorption spectrum and the time evolution on dissociative pathways. Other methods enable very large scale calculations of the structure of complex molecules and of the potential energy surfaces on which reactions occur. Quantal flux propagation promises to unveil the motion through curve crossings unfettered by the problems of visualizing the interferences between ingoing and outgoing quantum mechanical waves.

Quantum-state-resolved studies of atomic and molecular interactions have opened the door to exciting new possibilities. The close connection between theory and experiment will yield significant advances, with each area stimulating the other. The machinery for these fundamental studies is becoming sophisticated enough to allow modeling of practical systems, some of which urgently require a thorough understanding. Better state and time resolution will incisively test theoretical models and validate their application to ever more complex systems, many with important practical applications and impact.

Clusters

Clusters are chemical aggregates containing specific numbers of atoms and molecules. Understanding the physics of molecular and atomic clustered species provides the fundamental link between the properties of isolated species and condensed matter. Indeed, the unusual characteristics of clustered species often suggest their role as an additional state of matter, with interesting consequences for catalysis, atmospheric chemistry, and materials. Cluster science uses the tools of AMO research to explore the appearance, with increasing size, of bulk physical and chemical properties. It provides new insights into solid-state physics and has resulted in the discovery of entire new classes of large molecules, as related, for example, in the "buckyball" story in Chapter 1.

In the simple limit, two molecules may be joined by weak van der Waals forces in a low-temperature jet expansion. The study of these weak forces and the often unexpected geometries that ensue is one of the most productive areas of molecular spectroscopy today. The results demonstrate the nature of intermolecular forces in unprecedented detail and have broad implications even for the hydrogen bonding of large biological molecules. Methods for the creation of ever-expanding classes of chemical clusters have evolved using laser ablation sources, jets, and clever chemical kinetics.

It is now possible to study a solvated electron in size-selected clusters and to manufacture metal-carbon clusters of preferential sizes for use as novel reagents. The caging effect of two molecules surrounded by a solvent shell has been studied in detail with ultrafast laser dynamics. Aerosol and ice clusters have been discovered to be important in the stratospheric ozone reduction problem, challenging scientists to devise new methods to study heterogeneous kinetic processes.

Clusters of metals and semiconductors have properties that differ from either the small-molecule or solid-state limits. In simple alkalis such as sodium, collective plasmon optical resonances have been observed in clusters of just a few atoms. In contrast, the bulk ionization potential is approached only in very large clusters; thus there are both quantum and classical electrostatic size-dependent terms. Highly symmetrical physical structures have been inferred for large metallic clusters from regular patterns of gas phase reaction and chemisorption. In addition, mobility studies show a clear separation of clusters that are linear, spherical, or ring shaped. In simple free-electron metals, the appearance of closed shells of electrons leads to "magic numbers" in stability. Ten-atom clusters of magnetic elements have magnetic properties stronger and functionally different from those of the bulk material—the approach to bulk magnetism is only just beginning to be understood. The many unusual effects of clusters have stimulated substantial theoretical efforts, resulting in large-scale molecular dynamical simulations and the demand for more powerful computational tools.

Semiconductor crystallites, nanocrystals, or quantum dots 2 to 5 nanometers (nm; 10^{-9} meters) in diameter have band gaps that increase with decreasing size, due to a large electronic quantum size effect. Their optical and electrical properties can be "tuned" by size. Macroscopic amounts of nanocrystals with narrow size distributions and surfaces stabilized by chemical bonding to organic molecules have been produced. Smaller semiconductor clusters (tens of atoms) have structures and electronic properties that are only poorly understood at present. The surfaces of these small clusters are less reactive than the surfaces of the bulk materials. The bulk "unit cell" appears to develop quite slowly with increasing size. With the rapidly advancing understanding of cluster species, the potential to use them in electronics and other devices remains a highly attractive goal.

Physics of Nonlinear Optics

The study of nonlinear phenomena in optics originated with the development of the laser and has resulted in a number of exciting discoveries. One nonlinear phenomenon of great interest is the soliton, a wave that propagates without distortion. Soliton pulses ranging from several hundreds of femtoseconds to a few picoseconds have been generated in optical fibers. The distortion-less propagation of such pulses makes them of obvious interest for optical communications, because they can, in theory, transmit digital data at higher rates and over longer distances than is currently possible. Solitons in fibers have been realized using amplification and nonlinearity associated with stimulated Raman scattering. Solitons have also been generated in erbium-doped fiber lasers combined with undoped fibers. So-called dark solitons—short dips in an otherwise continuous beam of light—are of interest for optical switching.

Other nonlinear methods are being studied to generate coherent radiation. Lasers without inversion, that is, without the need for an auxiliary pumping process to produce a population inversion, have been investigated in the past with Raman interactions and resonant frequency conversion involving only two atomic states. Recently, there has been a resurgence of interest in this area with the development of concepts involving multilevel near-resonant systems. Closely related is the use of intense laser fields to reduce absorption in resonantly enhanced nonlinear optical frequency conversion. it is reasonable to expect that novel methods for the generation of coherent radiation will continue to be devised in the future.

Nonlinear optical processes have also been employed to produce states of light that display distinctly quantum mechanical properties. In the quantum theory of a mode of the electromagnetic field, two in-quadrature components are defined that play the role of canonical variables. The uncertainty principle sets a limit on the accuracy with which both variables can be simultaneously measured, but it is of course possible to measure one with arbitrary accuracy at the expense of admitting large fluctuations in the other. Thus it is possible to produce "squeezed" states of the field in which one of these components has very small quantum fluctuations. This capability opens the possibility of encoding and decoding information using that component that is relatively noise-free. Squeezed states can be produced in a variety of nonlinear optical processes. Experiments with atomic beams and collections of two-level atoms in cavities, as well as three-wave mixing processes, have demonstrated noise reductions substantially below shot-noise levels. Squeezed states are of particular interest for the detection of gravitational waves, where the signal is so weak that it would ordinarily be masked by quantum noise. Applications are anticipated in many situations in which it is important to enhance signal-to-noise ratios, and various spectroscopic applications have also been proposed.

Experiments measuring polarization correlations in the two photons produced in a cascade atomic radiative decay have tested some of the most basic aspects of quantum theory vis-à-vis "realistic" local hidden variable theories. These experiments have upheld the so-called Bell inequalities predicted by quantum theory. However, there are still possible loopholes that prevent one from ruling out "realistic" theories with absolute certainty, and new experimental tests of quantum theory have been devised involving nonlinear optics. In these experiments a quasi-monochromatic beam of light is split by a nonlinear medium into two beams, each with a frequency half that of the original beam. Correlations in the positions at which the generated photons are detected, with fixed polarizations, replace the correlations in the polarizations of the emitted photons in the earlier, atomic cascade experiments. These correlations are also subject to Bell-type inequalities. In particular, quantum theory predicts a nonlocal type of interference in spatially separated photon channels, and these effects have been observed.

Progress continues to be made in the understanding of optical instabilities and chaos in lasers and nonlinear optical devices. During the past decade, it has been well established that chaos is by no means an unusual mode of behavior in such systems. Indeed, experiments in this area were among the first to confirm various "universal" routes to chaos predicted theoretically. The focus more recently has been on spatiotemporal chaos associated with variations in the field transverse to the direction of field propagation. Nonlinear and chaotic behavior in coupled diode lasers is also being studied, as are ways to control the onset of chaotic behavior. The same basic concepts of nonlinear dynamics are found in work on atoms and molecules in electric and magnetic fields and in the dynamics of lasers and nonlinear optical interactions.

Laser Cooling and Trapping

Laser cooling is a technique for transferring momentum, and therefore kinetic energy, from an atom or atomic ion to the photons of a laser beam, as described in Chapter 1. Although first proposed in 1975, and first demonstrated in 1978, laser cooling has already been used to produce what some news stories have called the coldest particles in the universe. The lowest temperatures, in three dimensions, that have been demonstrated thus far have been limited by the recoil of the atom from single-photon emission. This "recoil limit" is of the order of 1 mK for optical transitions. At this temperature, atomic velocities are of the order of 1 cm sec^{-1}, and the corresponding de Broglie wavelengths are several hundred nanometers. Accessing this temperature regime, coupled with the parallel development of methods for trapping atoms and ions by their interaction with external fields, has enabled new and far-reaching opportunities for scientific investigations and technological applications.

A fundamental limitation to high-accuracy atomic spectroscopy stems from the motion of the atoms. Laser cooling reduces the importance of all motional broadening mechanisms, including the Doppler effect and the effect of a finite interrogation time, allowing development of improved atomic clocks. The most thoroughly investigated approach to date uses a microwave transition between hyperfine levels of a cloud of laser-cooled and trapped atomic ions. The "atomic fountain" described in Chapter 1 provides another promising approach. An exotic variant of the fountain in which the atoms fall onto a parabolic mirror and are reflected back has been examined. This "atomic trampoline" may allow for indefinite interrogation times without significant perturbations to the energy levels of the atom. Several groups have also succeeded in trapping and cooling a single atomic ion and are attempting to realize an optical frequency standard that would have a potential accuracy, limited by the second-order Doppler effect, of 1 part in 10^{18}.

Dramatic improvements in many of the tests of fundamental laws and symmetries described above have been made possible through the use of laser cooling and trapping. For example, the mass of trapped antiprotons was recently compared to the mass of the proton via ion cyclotron resonance. This precise measurement is thought to be one of most sensitive tests of the CPT theorem for baryons, which says that any Lorentz-invariant field theory must be invariant under the combined operations of charge conjugation C, parity inversion P, and time reversal T. Laser-cooled ions have already been used in experiments to test local Lorentz invariance and possible nonlinear corrections to the Schrödinger equation with unprecedented sensitivity. Tests of general relativity, such as the gravitational red shift effect, will also become much more sensitive using the increased accuracy that cooling and trapping will bring to atomic clocks.

Another major area enabled by laser-cooled and trapped atoms involves quantum collective effects. When indistinguishable particles are cooled and compressed until their de Broglie wavelengths are comparable to the mean particle separation, the quantum statistics of the particles will affect the behavior of the entire system. If the particles have integral spin, they will obey Bose-Einstein statistics and may undergo a Bose-Einstein phase transition at low temperature in which a significant fraction of the particles can occupy the macroscopic ground state of the system. Researchers have come tantalizingly close to achieving the goal of observing this effect in a weakly interacting system in the case of magnetically trapped, spin-polarized hydrogen atoms. There are also efforts in several laboratories to laser cool either magnetically trapped lithium or cesium atoms to the necessary temperature and density conditions. These atoms are made up of an even number of spin-1/2 subatomic particles and therefore have integral spin as a composite particle. In these experiments, optical probing of the atoms may provide insight into the dynamics of the condensation process and a means of detecting the atoms in the Bose-condensed phase. It is still an open question whether these atoms remain a gas at absolute zero temperature or will coalesce into a solid or liquid state. It may also be possible to create a weakly interacting degenerate Fermi gas by laser cooling atoms with half-integral spin, which obey Fermi-Dirac statistics.

Laser cooling techniques are being used in conjunction with "particle optics," as discussed in Chapter 1, to produce atomic beams of unprecedented intensity. Lasers can collimate the atomic beam emerging from a source and focus it to dimensions of the order of several tens of micrometers. These intense atomic beams will find applications in atomic collision experiments as high-flux sources for loading atomic traps and may be useful for lithography and nanoengineering (Figure 2.2).

Experiments involving single trapped atomic ions have also permitted fundamental investigations of the nonclassical nature of photon statistics. By isolating the resonance fluorescence from a single laser-cooled and trapped ion, experimentalists were able to observe electron "quantum jumps" between energy levels of the ion. Also, by recording the correlation functions between the photons emitted by a single emitter, the nonclassical effects of "photon-antibunching" and sub-Poissonian photon statistics were observed.

The observation of "Coulomb clusters" is an example of the far-reaching impact of laser cooling and trapping techniques in other areas of physics. E.P. Wigner predicted that a nonneutral plasma would undergo a phase transition from a gas to a liquid or a solid when the ratio of the Coulomb potential energy to the kinetic energy of the gas is increased to a value of approximately 100.

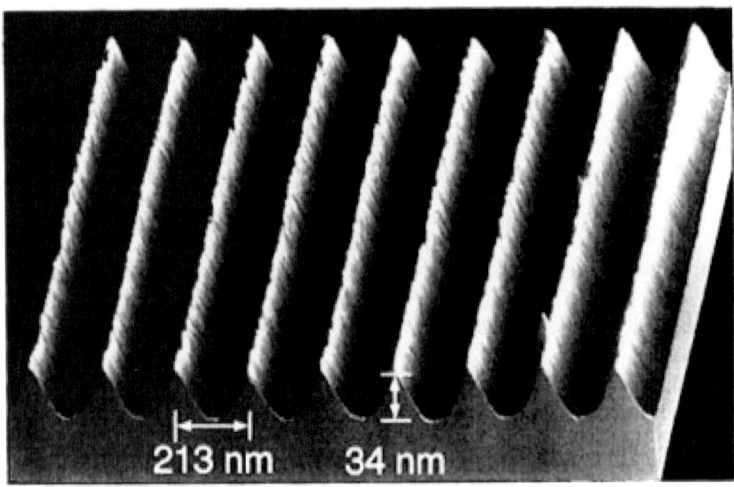

FIGURE 2.2 This atomic force microscope image shows "nanowires" of chromium laid down by laser cooling and focusing of an atomic beam. Such research is leading the way toward efficient methods for "nanofabrication" of what were previously impossibly small devices. (Reprinted, by permission, from J.J. McClelland, R.E. Scholten, E.C. Palm, and R.J. Celotta, "Laser Focused Atomic Deposition," Science **262** (November 5), 877-880, 1993. Copyright © 1993 by the AAAS.)

Although many attempts had been made to achieve the required conditions, it was not until laser cooling was brought to bear on the problem that success was achieved (Figure 2.3). Stable "solid" structures were observed with between 2 and approximately 30 ions. These "clusters" assume shapes that minimize the free energy of the system. Larger clouds containing hundreds of thousands of ions have been observed to form concentric rings in which the ions within a ring exhibit liquid behavior while the rings themselves remain quite stable.

Interactions with Surfaces

Interactions of atoms, molecules, and ions at surfaces of liquids and solids have long withstood attempts at detailed understanding, but new technologies—both experimental and theoretical—have opened the door to dramatic advances during the past decade. Much of this progress has resulted from developments in atom-ion beam technology and from advances in ultrafast lasers. Recent work has provided detailed information about atomic and molecular interactions with surfaces, information that is being exploited to improve the understanding of heterogeneous catalysis and to implement new technologies such as X-ray lasers and fusion reactors and to develop novel methods for the thin film deposition and semiconductor etching important for device fabrication. Ultrafast laser techniques have been applied to monitor the relaxation rate of excited adsorbates and to study charge transfer processes at electrode surfaces and between photoexcited adsorbates and the surface. Such information is crucial to the development of new photochemical synthetic methods at interfaces and new lithographic techniques.

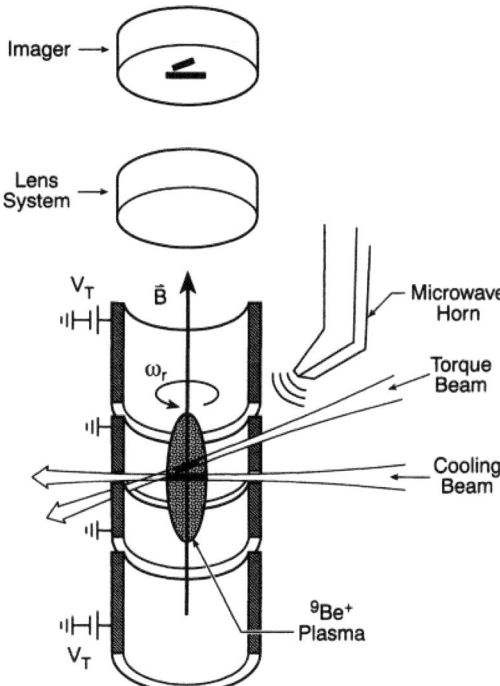

FIGURE 2.3 The trapping and cooling of ions in an electromagnetic trap has recently been used in many important advances. One of the most important is the trapping of a single ion and its interrogation, which may result in another advance in time and frequency determination. In the illustration here, the trap and lasers are used to cool ions to the point that they form "ion crystals," which may be viewed as a new form of matter. (Courtesy of David J. Wineland, National Institute of Standards and Technology.)

Surface studies, especially those at the quantum-state-resolved level, provide new challenges for theory in an area that lies at the interface between atomic and molecular theory and condensed matter theory. Questions concerning the nature of molecular adsorption and the dynamics of particle-surface interactions remain to be answered.

Studies of particle-surface interactions have resulted in the observation of new atomic species, including so-called "hollow atoms." These are formed by resonant neutralization of slow, highly charged ions incident at grazing angles on a metal surface and comprise essentially neutral atoms having most of their electrons in high-lying Rydberg levels. These highly excited atoms are very short lived and decay by Auger processes. The successive filling of inner shells is a true dynamical many-body process for which only classical statistical models are currently available. For large angles of incidence the incident ions penetrate the surface, and effects due to screening by metallic electrons become important, allowing study of the behavior of the excited states in the presence of screening. Basic questions remain to be answered, however, concerning the role of electron correlation in the formation of hollow atoms, the positions and widths of multiply excited resonances, and the nature of atomic states in the vicinity of a surface.

The study of arrays of atoms and molecules adsorbed on the surface of atomic or molecular liquids and solids is important for many applications. The continued development of electron spectroscopic techniques, often facilitated by the ready availability of synchrotron radiation, has allowed these structures to be probed at a level of detail never before possible. New optical spectroscopies, such as second-harmonic generation, have probed the liquid-solid interface at a molecular level for the first time. Many of these investigations have been directed toward establishing direct correlations between the chemical and physical information obtained from laboratory experiments carried out under single-collision, ideal conditions on crystalline surfaces and the corresponding reactivity in practical systems involved in heterogeneous catalysis. They have resulted in substantial new insights into catalytic reaction mechanisms, which have the potential to form the basis for improvements in catalyst performance. The emphasis is shifting from analysis of surface structure and properties to their control.

A profound development in the area of surface studies is the scanning tunneling microscope (STM) and the related atomic force microscope (AFM), which can image a surface with atomic resolution, thereby enabling the position (and orientation) of adsorbed atoms and molecules to be determined. These new instruments have been used to identify the active site on which a surface chemical reaction occurs. Such microscopic information is crucial to the formulation of a more general theory for catalysis. Tunneling microscopy has also been used to measure the electronic spectrum of single adsorbed molecules and to promote their reaction. Recent research has demonstrated that the position of an adsorbed atom or molecule can be manipulated using an STM and its derivatives. This may allow the building of intricate nanostructures on a surface and provide unparalleled control of surface structure and composition.

Enabling Other Fields of Science

The importance and impact of AMO science are in significant measure due to the critical role it plays in enabling many other scientific disciplines. Experimental techniques and measurement procedures that derive from AMO studies are widely used in other areas of science, including astrophysics, space science, atmospheric and environmental science, plasma physics, exotic atoms and nuclear physics, surface and condensed matter physics, and the biosciences. Information regarding the properties of atoms, ions, molecules, electrons, and photons is critical to modeling and understanding many physical environments.

Astrophysics

Apart from cosmic rays, all the information we have about the universe beyond the solar system is brought to us by photons. Astronomy relies on the interpretation of the distribution in frequency and intensity of the photons that are emitted by astronomical objects and detected by ground- or space-based telescopes. Even information about the earliest stages in the history of the universe before any nuclei, atoms, or molecules existed, before the galaxies and stars had formed, is carried to us by photons. The differential microwave radiometer on the Cosmic Background Explorer satellite tells us about conditions in the early universe by measuring the photon intensities at millimeter wavelengths.

The processes that create the photons and modify them in their journey to Earth through intergalactic and interstellar space from their distant origins belong to the domain of AMO science, as do the instruments that monitor these photons and measure their spectra. The photon distribution with wavelength is used to define the kind of object and to distinguish among different kinds of galaxies, stars, pulsars, jets, stellar winds, supernova and supernova remnants, nebulae, masers, protostars, and molecular clouds. A major triumph of astronomy has been the placing of stars in an evolutionary sequence constructed from analysis of stellar spectra that change as stars are born, grow old, and die.

To achieve a comparable understanding of other astronomical entities requires a comprehensive grasp of AMO science supported by an extensive database of atomic, molecular, and optical parameters. Providing these data is a major challenge. Experimental laboratory studies are essential for high precision and to test theoretical procedures, but the huge and growing volume of data that are needed can be provided only by theoretical calculations. Data are not enough. A deeper appreciation of atomic, molecular, and optical processes as they occur in the exotic conditions found in astronomical environments is also essential and can be gained only by basic experimental and theoretical AMO studies. We can also learn about atomic, molecular, and optical properties from investigations of astronomical phenomena. Despite the enormous differences in spatial and time scales, the fundamental processes in laboratory and astronomical plasmas are the same.

Lasers are becoming increasingly important in astronomical measurements. They are used to accurately measure and control baselines in coherent interferometry in the microwave region. Dust grain distributions around stars have been measured by using a carbon dioxide laser as the local oscillator for heterodyne interferometry. Laser-produced artificial high-altitude light sources have been used recently in adaptive optical wavefront correcting systems. The image quality in ground-based astronomy has been severely limited by the wavefront distortion introduced by atmospheric turbulence. If a test wave is available, this distortion can be measured and corrected for by using a deformable optic allowing large ground-based telescopes to have resolution similar to that of comparable space-based systems. Over a limited range of the sky, high-magnitude stars can be used to provide the test wave. For most observations, however, the test wave must be generated by using laser scattering from aerosols in the lower atmosphere or fluorescence from sodium atoms in the mesosphere pumped by a laser source of sodium resonance radiation. Maximizing the sodium fluorescence requires a detailed understanding of the dynamics of the atomic interaction with the laser radiation (Figure 2.4). Hyperfine splitting, Doppler broadening, coherent excitation, saturation, optical pumping, and radiation pressure are all important.

In the next decade, new windows will be opened through which the universe will be observed. Ground-based and satellite-borne telescopes supported by powerful instruments for photon detection will provide a vast array of spectral data at high resolution in the radio, millimeter, submillimeter, infrared, optical, ultraviolet, and X-ray regions of the electromagnetic spectrum. The increased sophistication and detail of theoretical models of astronomical events and the

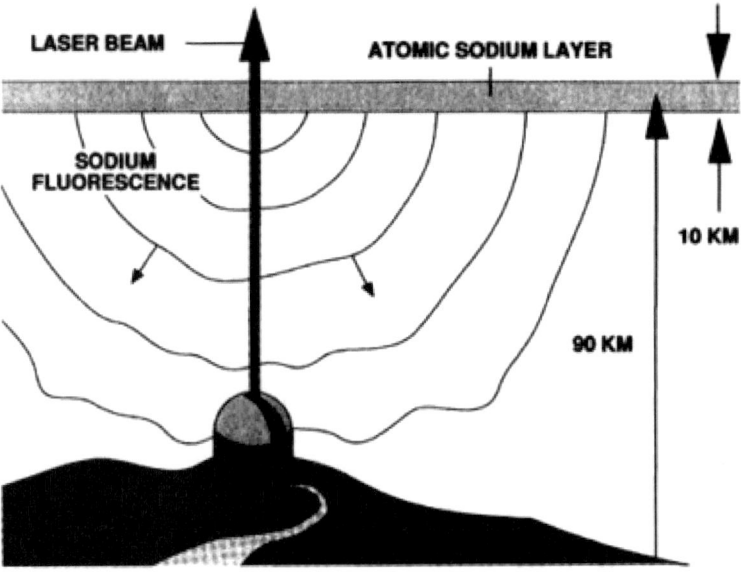

FIGURE 2.4 Mesospheric sodium beacon for atmospheric adaptive optics. (Reprinted with permission of MIT Lincoln Laboratory, Lexington, Mass.)

quality of the observational data will place severe demands on AMO science that must be met if the full potential of astrophysical research is to be realized.

The central role of AMO science is evident in a broad range of significant astronomical questions. For example, the helium abundance in the primeval universe constrains the number of neutrino flavors and the age of the universe. The ratio of deuterium to hydrogen (D/H) provides a measure of the baryon density in the early universe. This ratio can be obtained directly in the solar neighborhood from the absorption of starlight by deuterium and hydrogen atoms. To determine the D/H ratio over wider regions of the galaxy, measurements of deuterated molecules like DCO^+ and DCN can be used. Deuterated molecules undergo extensive chemical fractionation that enhances their abundances. To extract the D/H ratio requires detailed models of the fractionation, but the D/H ratio inferred from the observations is consistent with the value obtained for the solar neighborhood, suggesting it is a characteristic of the entire galaxy.

The D/H ratio in the universe is thought to diminish in time because of nuclear burning. Observations of DCO^+ in external galaxies coupled with a precise description of the atomic and molecular processes that lead to fractionation would test the assumption that the D/H ratio decreases with time. If it does, the D/H ratio measured for the galaxy establishes that baryonic matter cannot close the universe. It is possible that with improved infrared technology the deuterated hydrogen molecule could be seen in external galaxies and provide an additional measure of the D/H ratio, but a sophisticated interpretation of the emission would be necessary.

Recently, substantial advances have been made in building a theoretical description of molecular processes in interstellar clouds to the point where the molecular composition can serve as a chemical clock for determining the age of a cloud so that clouds can be placed into an evolutionary sequence. The changes in chemical composition that occur during the birth of a star have been explored, and diagnostic probes have been worked out in which the emission line widths of specific molecular species are used to investigate the phases of gravitational collapse occurring deep inside molecular clouds and hidden from view except through molecular emission lines at millimeter and radio wavelengths. To carry through detailed analyses of any reliability, a comprehensive set of molecular reaction rate coefficients at temperatures ranging from 5 to 2,000 K is needed. Greatly improved data on X-ray spectra are expected to flow from spacecraft launched in the next several years. Except in a few cases, the collision cross sections, recombination rate coefficients, and radiative transition probabilities are currently available only as estimates of limited reliability.

As a final example, supernova SN 1987a, the brightest supernova observed since the invention of the telescope, has been of enormous importance in astronomy. Observations over the years since the progenitor star exploded have yielded a store of information about stellar evolution, explosive nucleosynthesis, and stellar dynamics, all obtained by analyses of the spectra. After the first few days, the supernova became in effect a dynamically evolving laboratory of atomic and molecular physics, and the interpretation of the data has raised many questions in atomic and molecular physics that have to be addressed before reliable conclusions can be drawn about the dynamical evolution of the ejecta and the interactions with the interstellar medium in which the explosion occurred.

Space Science

In the context of this report, space science refers to the study of the solar system. It considers the interplanetary medium, the planets and their atmospheres, and comets, asteroids, and small bodies. AMO science provides the atomic and molecular energy levels, optical transition probabilities, cross sections, and rate coefficients that are the central parameters in quantitative models.

A recent example of the important role of AMO science in space is the discovery of the molecular ion H_3^+ and the explanation of unidentified lines in the Jovian aurora. Theoretical calculations of the potential energy surface and the resulting energy levels, supported by laboratory studies of selected transitions, provided detailed predictions of the emission spectra of H_3^+. These corresponded to observed, but previously unidentified, Jovian infrared emissions, and these emissions from H_3^+ are now used to provide images of the Jovian aurorae.

Comets may carry the imprint of their origin in the interstellar medium prior to the formation of the solar system 4.5 billion years ago. The analysis of cometary data is based on elaborate models of the effects of atomic and molecular processes and solar radiation on material released as the comet more closely approaches the sun.

The study of planetary atmospheres and the search for the reasons why the atmospheres of the planets and their satellite moons have evolved in different ways provide insight into the mechanisms that operate in Earth's atmosphere and assist in developing an understanding of such critical issues as the warming of the atmosphere due to the increase in carbon dioxide and the destruction of ozone due to the industrial release of chlorofluorocarbons. These processes are discussed further in the next section. Comparisons of the atmospheres of Earth, Mars, and Venus demonstrate the sensitivity of atmospheric composition to perturbations.

The escape of gases from the planets into interplanetary space has a major influence on atmospheric composition. Escape can occur through a variety of atomic and molecular processes. On Mars, nitrogen loss is driven by dissociative recombination; on Venus, helium loss is driven by charge transfer. The ionospheres of all the planets are caused by the absorption of solar radiation in photoionizing events and by ionization through solar cosmic ray impacts. Aurorae are a feature of planets in which excitation is caused by the impact of fast electrons and ions, accelerated in the magnetosphere. The atmospheres respond to radiation with a great variety of molecular reactions, and the construction of realistic models of atmospheric behavior is impeded by the lack of an accurate database and also by an insufficient grasp of the enormous variety of atomic and molecular processes that occur.

Space physics differs from astrophysics in that local measurements are possible using spacecraft. Satellite-borne instrumentation has yielded basic data on planets, comets, and the magnetosphere. This instrumentation is in many cases directly descended from apparatus developed for laboratory-based AMO studies.

Doppler shifts and line widths of molecules in space and planetary atmospheres have been measured using heterodyne interferometry. Carbon dioxide and lead-salt lasers have been used to detect carbon dioxide in the Martian atmosphere. Studies have been made of atmospheric wind variations, and information on altitude dependence has been obtained from observations of the line profiles. Laser sounding systems are also being developed as onboard instruments for planetary probes.

Atmospheric and Environmental Science

There is national awareness of the importance of protecting Earth's atmosphere, as reflected, in particular, in concern with global warming and with depletion of Earth's ozone shield. Global warming is the predicted change in the global temperature resulting from a steadily increasing level in the atmosphere of carbon dioxide and several other trace gases from anthropogenic sources. Although the gaseous contaminants occur in small amounts, their role in absorbing and radiating long-wavelength infrared energy has the potential to effect large changes in Earth's temperature. The actual changes that might occur are governed by a complex series of AMO processes. A similar situation holds in the case of ozone depletion in the stratosphere. A thin shield of ozone molecules protects life on Earth's surface from the destructive effect of solar ultraviolet radiation. There is evidence that anthropogenic sources are affecting the level of ozone in the stratosphere. Limiting these sources of damaging chemicals significantly affects the economies of both the United States and other countries. Thus an accurate understanding of the development of this problem has major implications for society and national economic planning.

Research directed at understanding our natural environment proceeds along three parallel fronts, in each of which AMO science plays a key role. The first involves the development and utilization of computer models to analyze the effect of anthropogenic physical and chemical perturbations and to predict the response of environmental systems to possible future variations in these perturbations. The second research front focuses on laboratory studies designed to isolate and characterize individual chemical and physical processes and to accurately quantify their controlling parameters. This includes measuring quantities such as rate constants, cross sections, line shapes, scattering and extinction albedos, and other basic physical parameters that are at the core of AMO physics and closely related fields. Of course, not all of the relevant microscale parameters are determined by experiments. Theoretical molecular structure and molecular dynamics studies can often yield valuable information, particularly for species and/or physical conditions that pose experimental difficulties. The third research front is the development of sensitive physical and chemical sensors and their use in field measurements to quantify the current state of critical environmental parameters such as trace species concentrations, temperatures, and phase distributions. Measurements are also essential for characterizing and quantifying rates of change in key parameters such as fluxes due to heat and mass transfer, chemical reaction, evaporation, nucleation, condensation, and solidification. AMO science is key to many atmospheric sensing methods.

The three lines of environmental research described above are closely linked. Laboratory research supplies critical information (atomic and molecular parameters) needed to refine and extend the computer models. It also provides field instrument developers with the information they require to quantify advanced sensor performance. Sensitivity analyses of model results can, in turn, identify the microscale parameters that most need further investigation. Models are also used both to design and to analyze the data from field measurement programs. The environmental quantities obtained in field measurements, such as the temporal and spatial variations of trace species concentrations or temperatures, provide the yardstick to judge a model's accuracy and predictive capability.

The trace species content of the global atmosphere is changing at a rapid rate, particularly compared to geologic time scales. Numerical computer models that attempt to quantify the flux of trace species into and out of the atmosphere, the fluxes of infrared, visible, and ultraviolet light through the atmosphere, the rates of atmospheric chemical transformation and physical state change (such as cloud formation, evaporation, and precipitation), and the transport of atmospheric mass (such as by wind or diffusion) are the repository of all we know quantitatively about atmospheric model assessments.

The most sophisticated of these models rely on a wealth of AMO physical detail. For instance, as sunlight penetrates the atmosphere, its ultraviolet component initiates critical photochemical processes by photodissociating O_2, O_3, HO, NO, ClO, NO_3, and a variety of other atmospheric species. Ultraviolet and visible light is also scattered by cloud droplets, clear air aerosols, and Earth's surface. All of these absorption and scattering processes depend strongly on the wavelength of the light and the local chemical and physical state of the atmosphere. Models of the atmospheric transport of solar radiation require sophisticated theoretical techniques to allow adequate representation of the effects of physical optics such as scattering and absorption by atmospheric droplets and particles. These scattering models are combined with the quantitative molecular spectroscopy necessary to represent the effect of gaseous species in order to accurately reproduce the temperature field of the upper atmosphere and the photochemical driving force for chemical change in both the upper and the lower atmosphere. However, even the most sophisticated solar radiation transport model is only half the atmospheric radiation transport story. The outgoing "earthshine" at infrared wavelengths is critical to temperature fields at Earth's surface and in the lower atmosphere. A full model must also adequately represent the absorption due to at least a half a dozen key infrared active molecular trace species, all with different altitude profiles and with pressure-sensitive infrared line shapes. Furthermore, the ability of clouds of different types to absorb and re-emit infrared radiation depends critically on their altitude, optical depth, cloud droplet size spectrum, and physical state as well as their temperature relative to that of Earth's surface and other cloud layers.

Once an adequate representation of radiation transport has been achieved, chemical transformations in both the gas and the condensed phases (due to heterogeneous interactions between trace gas species and cloud droplets and/or aerosols) must be considered. Models that include several hundred homogeneous gas-phase and heterogeneous condensed-phase reactions are no longer unusual. In addition, heat and mass transport, both within the atmosphere and at the lower (Earth) and upper atmospheric boundary layers must be properly parameterized. Finally, modelers may add in the nucleation and growth of aerosol particles from high-vapor-pressure substances produced by atmospheric chemistry, which can, in turn, serve as condensation sites for cloud droplets and ice particles, leading to substantial perturbations on both heat flux and mass transport in and around the cloud and significant chemical perturbations via heterogeneous reactions. It is the balance of all these AMO processes that determines the progression of global warming under the influence of the greenhouse effect. In the same way, a complicated photochemistry determines the ozone depletion in the stratosphere.

The tools, techniques, and data provided by AMO science are of vital importance in addressing global atmospheric environmental problems such as stratospheric ozone depletion, greenhouse gas buildup, and acid deposition and tropospheric oxidative capacity. Atmospheric modeling and the interpretation of atmospheric diagnostics require accurate knowledge of molecular and atomic energy levels, transition probabilities, photodissociation and photoionization cross sections, reaction rates, and collision cross sections. The available data are incomplete, especially as concerns infrared and ultraviolet transitions in trace molecules and free radicals and reaction rates involving free radicals. An expanded database is essential to advances in atmospheric modeling for meteorology and the prediction of climate change and the effect of local and global pollutants.

Plasma Physics

AMO science plays a pivotal role in the study of plasmas. The formation and evolution of laboratory plasmas depend on microscopic processes that influence the material equation of state and determine the energy production and transport. Also, atomic and molecular spectroscopic measurements provide a direct probe of the plasma environment, yielding its temperature, density, velocity, and internal electromagnetic field strengths. The quest for inertial and magnetic confinement thermonuclear fusion demands an extensive database describing atomic phenomena in high-temperature plasmas in which intense electromagnetic fields and shock waves may be present. Conventional energy production devices, such as magnetohydrodynamic generators, have spurred extensions of our understanding of complex atomic interactions in external electromagnetic fields. Modern defense programs rely increasingly on our knowledge of high-energy-density plasmas. In laboratory X-ray laser studies, our ability to model the phenomena satisfactorily is sometimes limited by an incomplete understanding of subtle quantum electrodynamic effects on the shapes of spectral lines. Low-temperature plasmas are fundamental to commercial and residential lighting. Plasma processing of materials is of vital importance to several of the largest manufacturing industries. It is indispensable for manufacturing the very-large-scale integrated microelectronic circuits used in computers, communication equipment, and consumer electronics. Plasma processing of materials is important in the automotive, aerospace, steel, and biomedical industries. A severe limitation in the design of plasmas for material processing is the lack of a detailed understanding of the atomic and molecular processes that occur. Advances in technology are providing new capabilities for the study of atomic processes in plasmas. Magnetic confinement fusion experiments can attain long-lived plasmas at temperatures of tens of kilovolts through the careful control of atomic processes involving impurity ions and the interactions of the bulk hydrogen plasma with the walls of the device. Large laser facilities have compressed target materials to hundreds of times their normal solid densities, reaching conditions heretofore found only in astrophysical objects. In addition to the fundamental attraction of producing atoms in strong electromagnetic fields, short-pulse lasers offer opportunities to study warm, dense plasmas without the complications of hydrodynamic motion that inevitably occur on longer time scales.

The advent of supercomputers has permitted theorists to make more accurate calculations of individual processes as well as integrate many atomic processes into comprehensive models of plasma behavior. In several instances, plasma modeling has progressed from a set of descriptive methods for explaining crude observations to a set of procedures for the design of practical devices. A full understanding of how and to what extent a plasma environment actually modifies reaction rates is lacking. Indeed, after much study, even for the relatively simple processes of direct collisional excitation or ionization, substantial disagreement still remains between rate coefficients measured in a plasma and those calculated from isolated atom cross-section data.

Complications arise because the importance of a particular process in a plasma is affected by the state of the plasma itself. For example, the rate at which collisional ionization of a given ionic species occurs is determined by the populations of its excited levels, which in turn depend on the velocity distributions of plasma electrons and ions, or the ambient radiation, and sometimes on the presence of an intense electric or magnetic field. In dense plasmas, the simple picture of atoms in a plasma breaks down, and methods similar to those developed in condensed matter physics and the physics of liquids must be invoked.

Numerous opportunities and challenges for AMO science can be identified in plasma physics. At low temperatures, there is an immediate need for accurate measurements of the complex molecular processes that occur in partially ionized plasmas. This includes both surface phenomena and interactions by which complex molecular structures are formed in the plasma. Intense, short-pulse lasers offer the opportunity to study transient phenomena in warm, dense plasmas on time scales that are short in comparison with time scales for hydrodynamic motion and many plasma-cooling processes. High-energy-laser facilities will be capable of compressing and heating matter to conditions hitherto found only in astrophysical environments, thus also providing new insight into cosmic phenomena. The increasing availability of sophisticated computers will allow accurate calculations to be performed for a wide range of processes. The quality of plasma modeling will be enhanced and will provide the first estimates of the importance of processes that could not be included in previous models. New formulations will be developed to describe the complex many-body processes discovered in the past decade. As new fundamental ideas are developed, it will be necessary to provide workable approximations for use in practical applications. Of particular importance is the development of simple models that explicitly include the interaction of the atomic and plasma environments. New directions in computer design, especially of massively parallel computers, present the opportunity to revisit the very foundations of computational atomic and plasma theory.

Exotic Atoms and Nuclear Physics

AMO physics can probe directly the properties of nuclear matter. As discussed earlier, experiments to test fundamental symmetries such as parity nonconservation are often based on precision atomic and molecular spectroscopy. Their interpretation makes use of refined atomic and molecular many-body calculations whose accuracy can be assessed by comparison with other atomic and molecular properties. Atomic parity nonconservation experiments may provide a unique method to measure neutron distributions in heavy nuclei. Experiments on hyperfine structure are a major source of information on nuclear magnetic dipole and electric quadrupole moments, nuclear sizes, and deformations. They require sophisticated atomic calculations to derive quantitative information about nuclear properties.

Collisions at low energies between systems of high nuclear charge z, such as uranium nuclei, are a potential source of information about quasi-molecules with transient nuclear charges in excess of the inverse of the fine-structure constant and could provide a stringent test of quantum electrodynamics. Unexpected phenomena have already been found in the form of electron-positron pair production arising from the decay of neutral particles.

The study of exotic atoms and molecules involving the states of negative elementary particles bound to nuclei is a rich and continuing area of research, giving insight into nuclear structure and hadron interactions. In exotic atoms and molecules an electron is replaced by another negative elementary particle, the most common being the muon (μ^-), pion ($^-$), kaon (K^-), and antiproton (p^-). All except the antiproton are unstable but have lifetimes that are long in comparison with many interesting chemical time scales. The relatively long-lived muon is uniquely important in that, like the electron, it has no strong (hadronic) interactions and provides both fundamental and practical insights into the structure of nuclei.

Muonic atoms have been studied for many years, but modern facilities have made this work much more productive. Careful measurements of the energies of cascade X-rays have provided much information on the sizes and shapes of many nuclei. Even more information can be obtained by using polarized atoms. Recently, large polarization of the muonic helium atom (30%) has been achieved by adding a small amount of rubidium vapor to a gaseous helium target. The rubidium is laser polarized and in turn polarizes the electron in the ^{4}He-μ^--e^- system by exchange, which then polarizes the muon. This system will be used with ^{3}He to measure the recoil asymmetry of muon capture to determine the induced pseudoscalar coupling constant (this tests the partial conservation of axial current and our understanding of hadronic effects on the weak current).

Muons can also be used to catalyze fusion reactions. In this process, two hydrogenic nuclei, for example, a deuteron and a triton, are bound to form a hydrogenlike molecular ion but with the internuclear distance reduced by a factor of over 200 owing to the large mass of the muon (207 times greater than the mass of the electron). This suppression of the Coulomb barrier allows fusion to occur rapidly (on time scales of a picosecond) and presents a method to achieve nuclear fusion without the high-temperature and confinement difficulties inherent in plasma fusion concepts. However, although a yield of ~150 fusions per muon has been experimentally observed in deuterium-tritium mixtures, this is still well short of that required for energy production. The number of fusions is limited by the cycle rate (compared to the muon lifetime) as well as the probability that the muon "sticks" to the helium nucleus produced by nuclear fusion. The precise calculations of both the binding energy of the deuterium-tritium-muon molecule and the sticking probability have been challenging and have required the development of new theoretical techniques. Variational methods of unprecedented accuracy have been refined for the three-body problem involving long-range interactions and are expected to find application in other areas. There remains a significant disagreement between the theoretical and the experimental values of the sticking fraction whose resolution might reveal some important new physics.

The low-energy antiproton ring (LEAR) at the European Center for Nuclear Research (CERN) is now capable in principle of producing antiprotonic atoms, and a number of important experiments are anticipated. Because of the large mass of the antiproton, states with extremely high principal quantum numbers will be populated. Topics of interest include formation and cascade at extremely high principal quantum numbers, the effect of the strong interaction of the antiproton with the nucleus, annihilation processes, and static particle parameters such as the mass and magnetic moment of the antiproton determined by using atoms in sufficiently high states that the strong interaction is negligible.

The lifetime of an antiprotonic atom in a high Rydberg state is usually limited by internal or external Auger processes. A novel class of antiprotonic helium states has recently been discovered that suppresses Auger quenching. The system consists of ^{4}He-p$^-$-e$^-$ in a highly excited moleculelike configuration. In pure helium, these states have been observed to live for microseconds, much longer than the subpicosecond lifetimes normally associated with Auger processes. Further experiments are under way to detect visible light produced in the deexcitation process and to pump the metastable atoms with lasers, perhaps prolonging this lifetime even further. It may even prove possible to catalyze antihydrogen production via reactions of the metastable atoms with positrons or positronium.

Surface and Condensed Matter Physics

A rapidly expanding arsenal of analytical methods based on particle-surface interactions has been developed for use in surface and materials characterization. Techniques based on electron and atom diffraction are used to examine surface order and structure. Auger spectroscopy and ion backscattering are employed to determine surface composition and monitor contamination. Recently, a variety of spin-sensitive spectroscopies have been implemented that, coupled with advances in the technology to grow thin epitaxial films and engineer artificial structures, have made possible the observation of a variety of novel magnetic phenomena at surfaces and interfaces. One especially powerful technique developed to examine magnetic microstructure is scanning electron microscopy with polarization analysis (SEMPA), in which a highly focused electron beam is directed at the surface of interest and the polarization of the ejected secondary electrons is measured. This approach allows high-resolution studies of domain structure and boundaries with applications in magnetic recording (Figure 2.5). Small-angle ion-surface scattering also provides a powerful probe of surface magnetic properties. Using this approach, it has been shown that surface magnetic properties frequently differ from those of the underlying bulk and that short-range magnetic order persists even above the surface Curie temperature. Surface magnetic properties have also been examined by investigating spin dependencies in the interaction of thermal energy electron-spin-polarized rare-gas metastable atoms with surfaces. This technique is particularly surface-specific because the incident atoms do not penetrate the surface.

FIGURE 2.5 SEMPA image of magnetic domains in a patterned permalloy array. Black and white contrast correspond to magnetization pointing to the left and right, respectively, while the gray regions between the stripes are the silicon substrate. The permalloy stripes are 2 µm across. (Courtesy of J. Unguris, D.T. Pierce, and R.J. Celotta, National Institute of Standards and Technology.)

The advent of ultrashort pulse lasers and time-resolved optical measurement methods has contributed substantially to condensed matter physics and has enabled the direct study of dynamical phenomena in solids and at surfaces on time scales typical of the phenomena themselves. Accordingly, much new information has been obtained, primarily in the areas of surface dynamics, nonequilibrium heating of metals, carrier dynamics in semiconductors, and relaxation phenomena in amorphous materials.

Femtosecond lasers have played a key role in the area of surface dynamical processes. A long-standing question about whether surface processes such as desorption are primarily thermal or electronic in nature has been resolved by time-domain observations of the desorption event. The fast time scale on which desorption is observed to occur has revealed new desorption mechanisms and has stimulated interest in the role of the substrate nonequilibrium electronic properties in surface dynamical events. Also of interest is how the presence of a nearby surface affects the lifetimes of excited states of atoms and simple molecules. To that end, the lifetimes of image states at metal surfaces have been measured directly, and the extremely short lifetimes observed point to the importance of electronic deexcitation mechanisms in, for example, chemical processes occurring at surfaces.

The nonequilibrium electronic properties of bulk solids is an active research area. In the first few hundred femtoseconds after a laser pulse strikes a solid, the electronic temperature of the solid is highly elevated, but the solid remains vibrationally cold. In times corresponding to approximately the inverse of the highest relevant phonon frequency, the lattice temperature begins to rise, and subsequently the hot electrons reach thermal equilibrium with the solid. The distribution and cooling of hot electrons have been studied optically by femtosecond thermomodulation and, more directly, by observing photoelectron emission on this time scale. These studies have provided a new and direct method for measuring electron-phonon coupling constants in solids, important in understanding superconductivity. They have also resulted in the first detailed observations of thermalization dynamics in metals, directly yielding electron distributions during nonequilibrium cooling.

In semiconductors, carrier thermalization of highly excited electrons and holes has been studied extensively, but fundamental questions remain. The intraband thermalization via phonon emission has been elucidated in gallium arsenide, and electron-phonon interactions in the space-charge region near semiconductor surfaces have been investigated. Many-electron effects are important in the screening and initial thermalization of hot electrons and holes but are still poorly understood, and further theoretical and experimental efforts are required. In amorphous semiconductors and glasses, relaxation effects can occur on widely varying time scales, owing to electron localization effects and distributions of low-energy trap states. Time-resolved investigations of carrier relaxation in these systems have provided an important method for establishing the presence of mobility edges and can probe the interaction between localized and extended states in these materials.

Biosciences—Mapping the Human Genome

The development and utility of optical tweezers have been noted in Chapter 1. Here the focus is on the use of optical techniques for mapping the human genome, because these promise greatly increased sequencing rates. The human genome is composed of approximately 3×10^9 base pairs organized into 23 continuous subsequences (chromosomes). With overlap requirements (fragments are sequenced and overlap regions used to match fragments) and duplications necessary to remove ambiguities, it is estimated that about 5×10^{10} base pairs of raw sequence must be accumulated. In order to map the genome, it is necessary to develop a fast process for recording base pairs in DNA. An optical technique is being developed that will increase the raw sequencing rate by a factor of 1,000 over current, state-of-the-art automated sequencing techniques. The approach is to synthesize a complementary strand of DNA using fluorescently labeled nucleotides in the reaction mixture. Each of the four bases (A, C, T, and G) is tagged with a nucleotide having a different fluorescent dye. The fluorescently tagged DNA fragment is attached to a microsphere. The microsphere is placed in a flowing sample stream, and the individual, fluorescently labeled bases are cleaved from the DNA fragment sequentially by an oxonuclease and are identified by photon burst fluorescence as they pass downstream. The light forces associated with a focused laser beam are sufficient to hold and move individual microspheres and allow manipulation of individual DNA strands.

In another optical approach to the Human Genome Project, genome sequencing using direct imaging X-ray color holography is being explored. This technique is similar to fluorescence labeling, but the four bases in a segment of DNA are tagged with different heavy atoms, thus causing only small structural alterations to the configuration of the original (unaltered) bases. The segments are placed on a 100-angstrom (Å) carbon foil, and a tungsten sphere is positioned adjacent to the DNA segment. Irradiation with 5-Å radiation of sufficient coherence length can produce a holograph. Scattering from the tungsten sphere produces the reference beam, which interferes with the scattered beam from the DNA. A 1-micrometer coherence length is sufficient and can be obtained from an undulator. The scattered radiation (hologram) will be read by a two-dimensional, X-ray-sensitive charge-coupled device and the image of the DNA reconstructed.

THE NATION'S MEASUREMENT TECHNOLOGY

Among the ways a field of science can have importance are through its scientific interest, its impact on other fields of science, and its impact on technological or societal issues. The first two of these are briefly examined above. The third is examined here and in the following section, beginning with a look at measurement. The measurement standards and methods in use today are based primarily in AMO science, and a large fraction of modern measurement methodologies and instruments originate in AMO science.

Measurement Standards

Accurate and precisely interrelated measurements are essential to equity in trade, quality control in manufacture, access to the global marketplace, and the progress of science itself. In the second half of this century, we have witnessed a significant paradigm shift, led by AMO science, away from the traditional, hierarchical, and artifact standards to standards based on more sharply definable and independently realizable atomic and molecular parameters.

Earliest historical records indicate use of measures and standards primarily for equity in trade. For a local trading region and the limited accuracy required, the arbitrariness and inconstancy of the standards had little impact. However, with increasing world trade greater standardization was required, and this was provided by the Treaty of the Meter, which came into worldwide effect only a little over a century ago. This major advance in measurement science and standardization was, from a present-day viewpoint, flawed in its radical dependence on a singular collection of artifacts. Among these, one was designated as the basis for each quantity, as, for example, a particular meter bar and one of the kilogram replicas. These were (and in the case of the kilogram, still are) maintained under secure and stable condition at the Bureau des International Poids et Mesures (BIPM). The global measurement system was then realized by distribution of replicas, each normalized to the BIPM standards and periodically recalibrated. In the national standardizing laboratories such as the National Institute of Standards and Technology (NIST; formerly NBS) in the United States, these metrological tokens were further transferred to "working" standards against which subsidiary replicas used in commerce, science, and industry were calibrated and the results thus disseminated. This intricate structure met commercial, scientific, and industrial needs for a time but is no longer adequate.

In the second half of the twentieth century, a new paradigm has emerged, largely through AMO science. In this, the basic standards are connected as closely as possible to precisely measurable intrinsic atomic (or molecular) properties or to the fundamental constants of nature. The base units are then freely realizable by any laboratory or technical installation having the necessary resources. In practice, of course, not all measurements require ultimate precision and accuracy, so remnants of the hierarchical system remain in recommended realizations of secondary standards. There is a clear trend, however, to develop even the working standards as independently realizable systems, thus immune to the need for calibration by a central authority.

Up to the present, this shift in standards philosophy has been successfully applied and internationally accepted for all principal base units with the single exception of the unit of mass. Only the kilogram remains artifact-based and hierarchically disseminated. Even in the case of mass, however, AMO scientists are taking promising steps toward a possible atomic standard and replacement of this last vestige of artifact-based measurement. The prototype for the new atomic standards paradigm is the standard of time/frequency, now based on the oscillations of an undisturbed cesium atom. Time and frequency are the most accurately measurable of all physical quantities. Accuracies now achieved are better than 1 in 10^{13}, with precisions approaching 1 in 10^{17}. AMO scientists are conducting research that is anticipated to push the ultimate accuracies to parts in

10^{18} using laser manipulation and cooling techniques described earlier. In 1984 the unit of length was redefined in terms of a fixed (defined) value for the speed of light. Recommended realizations of the "length" definition through stabilized lasers permit length measurement accuracy of about 2 parts in 10^{10}, with length measurements thus becoming a subset of frequency metrology.

Measurement and Instrumentation

The continued advance of science and technology relies heavily on improvements in measurement techniques and instrumentation, and this is an area in which AMO science plays a key and facilitating role. Indeed, many discoveries have resulted directly from improvements in measurement techniques. The ability to observe the world via instruments that extend our limited powers of observation allows us to continually increase our understanding of the universe. In addressing questions of global change, AMO technology now makes possible precise measurement of toxic gases in the environment and of changes in the key components of Earth's atmosphere. In the world of the ultrasmall, the continued improvements in the performance and resolution of microscopes have driven a multitude of discoveries in the biological and materials sciences. In addition, these advances in microscopy have nurtured the development of very-large-scale integration of semiconductor devices on a single chip.

AMO science is central to the precise measurement of time, upon which the synchronization of our complex civilization depends. For example, it is now possible to localize one's position on Earth to an accuracy of a few meters by using atomic clock timing signals from navigational satellites, which may well lead to safer travel and could form the basis of automated highways.

As science advances, as society becomes more conscious of the way it cares for the environment, and as more sophisticated products enhance our lives, more refined measurements and more refined understanding of the phenomena attendant to the event or activity are being seen as necessary and useful capabilities. AMO science is the cornerstone in the infrastructure associated with these advances.

AMO science has in recent years produced a number of enabling technical advances that have led directly to new instrumentation and measurement technology. Some of these advances have resulted in commercial products; others have not yet reached that stage but are almost routinely used by researchers in virtually all fields of physical science and engineering throughout the nation and the world. A few of the more notable AMO technical advances include the following:

- Atom interferometry, which may lead to orders-of-magnitude improvement in applied areas such as navigation and geophysics.
- High-resolution optical spectroscopy based on lasers, including Raman and other nonlinear optical spectroscopic techniques, with applications ranging

from atmospheric remote sensing of pollutants or other constituents to noncontact temperature measurement in flames.
- Ion beam techniques for doping semiconductors, optical metrology for lithographic patterning, and optical surface diagnostic techniques, which are all essential parts of today's modern semiconductor integrated circuit industry.
- Ultrashort optical laser pulses for measurement of temporal phenomena that happen on time scales extending into the femtosecond regime with important applications, for example, in the fundamental study of vision and photosynthesis process as well as in modern medical procedures.
- Carefully controlled and measured pulsed optical power for use in controlled ablation and evaporation of materials. For example, high-temperature superconducting films are prepared by using high-energy laser pulses. Lower-power pulses scribe and trim integrated circuits and can weld tears in the retina or reshape the cornea.

AMO in Measurement and Sensing for Industry

One of the largest roles of AMO science in industrial applications has been in the area of measurement. On-line monitoring, process optimization, quality control, pollution control, and nondestructive testing simply would not exist without the ability to sense and measure the relevant quantities. For example, manufacture of low-loss, single-mode optical fibers begins with a glass preform that is grown from high-purity chemicals. On-line monitoring is needed during each manufacturing stage. The preform is then drawn into a fiber, and the tight dimensional tolerances required are maintained through real-time measurements with immediate feedback through a control loop.

As another example, commercial instruments combining gas chromatography followed by time-of-flight mass spectrometry of elutants have revolutionized the analysis of complex liquid mixtures of large molecules. AMO research and trained AMO scientists are essential in the invention of customized chemical analytical instruments such as the gas and liquid chromatographs, visible and infrared spectrometers, and mass spectrometers that are used in industry. The design and engineering of such instruments require a thorough training in molecular and optical science at all stages.

Robots in automated manufacturing must have "eyes" to sense their environment, and this input forms the basis for decisions and actions by the robot. The continual development of new improved eyes, or sensors, is one of the challenges in automated manufacturing, and it is no surprise that fiber optics and lasers have assumed prominent positions in modern sensor technology. Indeed, fiber-optic sensors have been developed to measure a wide range of physical observables and offer the advantages of high sensitivity, inherent immunity to electromagnetic noise, ease of remote placement, usefulness in hostile environments, and small size and low weight.

Increasing use of sensors, both optical and chemical, is critical for the development of automated manufacturing, which is widely seen as necessary for the renaissance of U.S. manufacturing in areas ranging from food and drink preparation to chemicals and high-tech electronics. Innovative sensor development represents an important opportunity for future research in AMO science.

THE NATION'S TECHNOLOGICAL INFRASTRUCTURE AND U.S. ECONOMIC PRODUCTIVITY, COMPETITIVE POSITION, AND SECURITY

AMO science is a significant contributor to the nation's overall technological infrastructure and to its economic health and security. Indeed, it is estimated in Chapter 5 of this report that the products of AMO science have a significant impact on well over 9% of the nation's GNP and are important to an even larger percentage. The many impacts of AMO science in these areas are illustrated by the following examples, which contain a number of recurring themes. For instance, the large amount of AMO data now available describing the properties of atoms, molecules, ions, and photons is essential to understanding and modeling many different environments and systems. The required database, however, is far from complete, and continued research is required to expand it to allow questions of current importance to be addressed. Lasers have enabled major advances in almost every aspect of science and technology. Nonetheless, their application is limited in many instances by barriers of cost and performance. To remove these restrictions, further research and development are required to produce a new generation of lasers that are more efficient, more reliable, and less expensive than those currently available, with wavelength and output characteristics tailored to specific applications.

Industrial Technology, Manufacturing, and Processing

AMO science is central to many industrial manufacturing processes and technologies. Today, manufacturing is a highly technical and automated business requiring advanced methods that must continue to improve and evolve to remain competitive. AMO science plays a substantial role in this evolution, and only with the aid of continued research and development can U.S. industries maintain a competitive edge and successfully compete in world markets. Contributions of AMO science to industrial productivity include advances in lasers in manufacturing, plasma processing of materials, and chemical manufacturing.

Lasers in Manufacturing

Lasers today find widespread application in industry to thermally treat, weld, cut, drill, mark, and trim material. One of the earliest applications, which is still being used, is laser scribing for parts identification and inventory. Laser beams can be focused to extremely small spot sizes, allowing localized application of thermal energy. This minimizes thermal distortion in objects during cutting or welding operations and permits highly accurate micromachining (Figure 2.6). With an appropriate choice of wavelength, lasers can be used to process a wide

FIGURE 2.6 Nd:YAG laser beam mills precise three-dimensional undamaged patterns in a lithium tantalate crystal. Gears and other complex mechanical parts can be made with characteristic diameters of the order of a few times the diameter of a human hair. (Reprinted, by permission, from Eli Wiener-Avnear, "Lasers Cut Microscopic Paths with Major Potential," *Laser Focus World* 29 (July), 75-80, 1993. Copyright © 1993 by PennWell Publishing Co.)

range of materials, including metals, ceramics, plastic, wood, and cloth, at rates that are frequently much higher than can be achieved with other methods. The ability to accurately control lasers with computers in real time makes them compatible with automated manufacturing facilities and allows for on-line design and processing changes. Although lasers are now common in industry, it is certain that their usage will increase because major improvements in the accuracy and efficiency of laser manufacturing processes can be achieved with further research to develop a better understanding of specific laser-material interactions and with the availability of lasers optimized for the particular interaction of interest. In many cases, lasers with higher reliability and lower cost are needed.

The use of lasers for microelectronic component and materials processing has advanced at a rapid rate over the past decade. Many of the important ideas and applications explored in industrial and university research laboratories have now found their way into the manufacturing environment. Other ideas are at an advanced stage of development and will become part of microelectronic technology in the near future. Integrated circuits (chips) today may comprise millions of microscopic transistors formed and interconnected on the surface of a semiconductor wafer and covering an area of approximately one square centimeter. The individual transistors and interconnects are formed through a complex sequence of growth and removal (etching) processes, many of which require submicron precision. Laser lithography systems are being developed for microelectronics manufacturing, because, in the ideal case, deposition or etching can be localized to the region where a laser impinges on the substrate, and local patterning is achieved without the use of masks or photolithographic process steps. Lasers also provide a means to repair or personalize circuits by cutting or fusing lines on chips, circuit boards, and modules. In the case of chips, this process is used to wire in new circuits or bypass broken elements.

In recent years, there has been considerable interest in the etching and ablation properties of a variety of polymers and ceramics processed in air using excimer lasers. The mechanisms that give rise to polymer ablation are complex and wavelength sensitive, but research has advanced to the point that it is now possible to utilize the ablation process of manufacturing, specifically to pattern polymer films. The primary polymer of present interest is polyimide, commonly used as the insulating dielectric on multilayered chip carriers or modules. Each layer of such a module requires thousands of blind holes typically a few tens of micrometers in diameter to interconnect metallurgy (generally copper conductors) on adjacent levels of the multilayered structure. A distinct advantage of the ablation technique for producing the blind holes is that the underlying metal layer is not damaged by the laser fluence used for ablating the polymer. For metal ablation, a much higher fluence is required. Thus, laser polymer ablation has a built-in etch-stop feature in this application, greatly simplifying manufacture. Other materials such as ferroelectrics and ceramics have also been successfully patterned using excimer laser pulses. Here, too, the advantage is the ability to carefully control the amount of material removed per laser pulse.

Laser ablation is also important in the production of new materials. A laser is used to ablate targets containing constituents that are collected on a variety of substrates. Under appropriate conditions, high-quality films are obtained that include, for example, high-temperature superconducting oxides, optoelectronic materials, and diamondlike carbon, materials of considerable importance to microelectronic circuits and devices. One high-temperature superconducting oxide of particular interest is $YBa_2Cu_3O_7$. High-quality films of this material have been grown on a variety of substrates (such as $SrTiO_3$). The advantages of laser deposition include simplicity, stoichiometric transfer of the neutrals and ions from target to substrate independent of laser fluence, and the need for only a relatively low thermal substrate bias temperature. This lower temperature greatly reduces undesired reactions between film and substrate. Superconducting thin films have numerous applications in, for example, thin film microwave devices, Josephson edge junctions, striplines (microstrip microwave circuits), and bolometers.

Plasma Processing of Materials

Plasma processing of materials is of vital importance in many manufacturing industries (Figure 2.7), most notably the electronics industry in which plasma processing techniques are indispensable for the manufacture of very-large-scale integrated (VLSI) microelectronic circuitry. Plasma processing is also a critical technology in the aerospace, automotive, steel, biomedical, and toxic waste management industries. Plasma processing technology is being used increasingly in the emerging technologies of diamond film and superconducting film growth. The area has been reviewed recently and is discussed in a recent National Research Council (NRC) report, *Plasma Processing of Materials: Scientific Opportunities and Technological Challenges* (National Academy Press, Washington, D.C., 1991). The 1991 report highlights the many important industrial applications of plasma processing. These are not all described here; rather, a few representative examples are given simply to illustrate the breadth of the applications.

Plasma-controlled anisotropic etching is a critical technology used in fabricating microelectronic devices (chips). It allows pattern transfer from a developed photoresist to the underlying structure without the undercutting that is characteristic of wet chemical etching. Plasma-deposited films of silicon nitride and silicon dioxide are important for many chip applications, including passivation and the insulating of different metal layers. Plasma deposition permits the use of lower substrate temperatures than would be required with alternate techniques, allowing the material to be deposited after completion of surface features that would be damaged by a higher processing temperatures. Plasma-enhanced chemical vapor deposition is used to grow amorphous silicon films for solar cells.

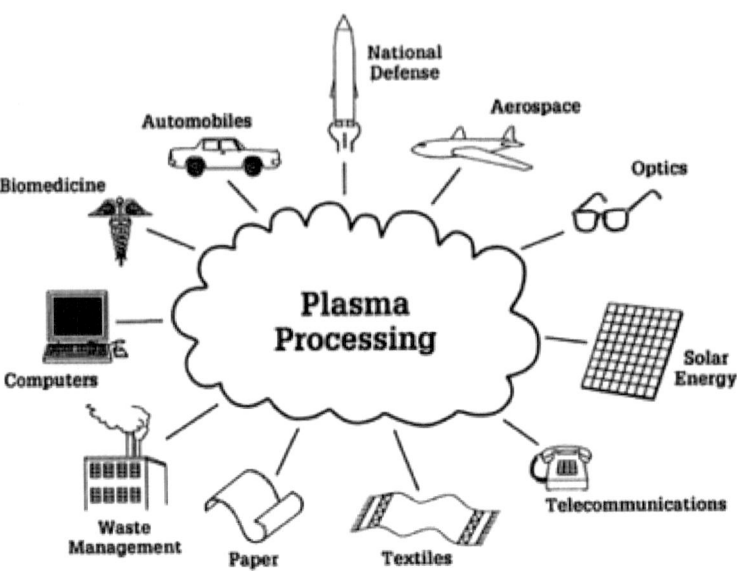

FIGURE 2.7 Plasma processing is a critical technology in many vital U.S. industries. (Reprinted from *Plasma Processing of Materials: Scientific Opportunities and Technological Challenges*, National Academy Press, Washington, D.C., 1991.)

Components of aircraft and automobile engines are protected against wear by plasma-spray deposition of ceramic or metal alloy coatings. Plasma-spray deposition and thermal plasma chemical vapor deposition are used to produce films of high-temperature superconductors and refractory metals. Plasma sputter deposition is important in the deposition of magnetic films for memory devices.

Despite the wide array of applications, plasma processing remains a largely empirical technology. The plasma processes in use today have been developed mostly by time-consuming, costly empirical exploration. The chemical and physical complexity of plasma-surface interactions has so far eluded the accurate numerical simulation that would enable process design. Similarly, plasma reactors have also been developed by trial and error. Nonetheless, fundamental AMO studies of surface interactions and plasma phenomena have contributed to process development by providing key insights into the operating conditions. These contributions include investigations of etching and deposition mechanisms using beams of reactive atoms, molecules, ions, electrons, and photons impinging on well-defined surfaces under controlled conditions, the development and use of diagnostic techniques such as mass spectrometry, optical actinometry, laser-induced fluorescence, and Raman spectroscopy to measure species concentrations in a plasma, and the measurement of rate constants and cross sections for reactions that occur in technologically important discharges. However, with further basis AMO research in these areas it should be possible to develop a more comprehensive understanding of the fundamental physical and chemical reactions that take place in plasma reactors. This, coupled with the recent availability of massive computational power, should advance the design of plasma reactors from an art to a science, further enhancing their effectiveness in manufacturing.

Chemical Manufacturing

Many AMO spectroscopic techniques, including atomic absorption spectroscopy, Fourier transform infrared spectroscopy, Raman spectrometry, and electron spin resonance spectrometry, are widely used in chemical manufacturing. Perhaps their predominant application is in the area of quality control and safety, and several representative examples are discussed here.

It is essential to protect consumers from toxic trace contaminants in either natural or manufactured products. Natural products may contain toxic elements that have been extracted from the soil and concentrated in the plant. For example, wheat selectively concentrates selenium, which is known to produce central nervous system damage at high concentrations. Although trace-level selenium is necessary for life, one can easily exceed the allowable limit. Atomic absorption spectroscopy is used to measure the selenium concentration. Traces of pesticide residues on fruits or vegetables must be eliminated, and only high-precision analytical spectroscopy (ultraviolet, visible, or infrared) can assure both distributors and users that the foods are safe. Another example of this type of problem concerns stannous fluoride (SnF_2), which is used as a fluoride additive in tooth-paste. Quite often, tin ores are contaminated with arsenic or other heavy metals, and these must be reduced well below parts-per-million levels before SnF_2 is used in dentifrices.

The accurate reproduction of color requires reliable, reproducible, and safe pigments and dyes. The quantitative evaluation of pigments, dyes, and other color sources requires not only a theoretical and an experimental understanding of atomic and molecular energy levels and transitions between them, but also a practical appreciation of the chemistry involved. Within the last few decades, the dangers of pigments containing lead and arsenic have been recognized. Dealing with this problem requires accurate analytical techniques. Modern photography requires a precise characterization of the dyes employed, in order to obtain a composite image that closely approximates the colors we see. The attractive colors in colored glass, synthetic rubies and sapphires, and various plastics are quantitatively characterized by use of visible-ultraviolet spectroscopic techniques allowing manufacture of reproducible products.

The pharmaceutical industry relies heavily on AMO science for the technology and instrumentation that it uses for quality control and assurance. Infrared, Raman, visible-ultraviolet, nuclear magnetic resonance, and mass spectrometry offer a broad range of approaches for analyzing the purity of samples and for characterizing their structure. The role of AMO science in precision measurement is important to this industry because many pharmaceuticals are specific isomers, and closely related molecular species are ineffective or even toxic.

New instrumentation and software have been developed in response to the increasing demands of federal and state environmental regulations. For example, inductively coupled plasma spectrometer systems have been produced that can measure 60 elements in less than one minute, using more than 5,000 emission lines to ensure interference-free analytical results. Environmental samples such as wastewater, soils, brines, and sludges can be rapidly analyzed by using such systems.

The discrete energy level structure of molecules offers the hope for selectively inducing chemical reactions using photons provided by, for example, flash lamps or tunable lasers. Selective photoexcitation of particular excited states opens the door to highly specific, mode selective chemical processing, whereby one can break a bond or lower an activation energy or make a desired reaction thermodynamically possible. As the availability of high-power lasers and flash lamps increases, it is to be expected that many applications of such synthetic chemistry will be discovered, including the laser-assisted deposition of diamond films from methane or halomethane molecules in hydrogen-rich environments.

Because of the many roles of AMO science in manufacturing, advances in the field can be expected to directly enhance economic vitality.

Information Technology, High-Performance Computing, and Communications

The success of industrial nations depends increasingly on their capability to develop and implement information technology. We live in the so-called Information Age, which is driven by the merging of communications and computing technologies. The foundations for the physical aspects of these industries lie in the AMO sciences. AMO research on laser sources and optical detectors as well as optical pulse propagation in fibers has provided the basic knowledge base for optical communications and optical data storage.

The status of and opportunities for these "photonic" technologies are discussed in the 1988 NRC report, *Photonics: Maintaining Competitiveness in the Information Era* (National Academy Press, Washington, D.C.). Two examples illustrate the role of AMO science in information technology—the erbium-doped fiber-optic amplifier and optical data storage.

The Erbium-Doped Fiber-Optic Amplifier

Telecommunications networks are rapidly evolving to an optical-fiber-based system that has higher performance and lower cost and requires lower maintenance than a copper-conductor-based network. Long-distance networks are now primarily fiber-optic, as are most links between local telephone switching offices. The last remaining link to the home is widely predicted to also become an optical link within the next 25 years. All of this progress is the direct result of advances in optical science, beginning with the invention of the semiconductor laser and subsequent research on fiber-optic transmission and fiber-optic systems.

One problem encountered in fiber-optic systems is that as the optical signal propagates through a fiber, it is attenuated due to losses in the fiber. To counteract these losses, expensive optoelectronic receivers and transmitter systems are normally used to reestablish the optical signals to usable amplitudes. These costly regenerators are restricted to a particular signal format and data rate and must be replaced whenever the system is upgraded to a higher transmission rate. Recent research, however, has produced erbium-doped fiber-optic amplifiers that eliminate the need for these regenerators. Such amplifiers not only boost (by a factor of 1,000) the signal back to its original level, but are also completely transparent to signal format or data rate (Figure 2.8). Furthermore, they can simultaneously amplify several different optical signals at slightly different wavelengths at the same time. Without fiber-optic amplifiers, a separate regenerator would be required for each wavelength channel.

One remarkable feature of erbium-doped fiber-optic amplifiers is the speed at which they have progressed from the research laboratory to application. The first efficient erbium-doped fiber-optic amplifier operating in the 1.5-micrometer (μm) region important for fiber communications was reported in 1988. Within 2 years, several major telecommunications research laboratories reported impressive fiber transmission system applications, including operation at 10 gigabits per second (10 billion bits of information per second) and simultaneous amplification of multi-optical-channel signals in a single fiber amplifier system. The fiber-optic amplifier has also been shown to be suitable for amplifying optical signals with various formats, including telephone, high-speed data, and television signals. Key to practical application was the parallel development of high-power semiconductor lasers operating at the correct wavelengths to act as power sources for the fiber-optic amplifiers.

It required only 4 years for the erbium-doped fiber-optic amplifier to make the transition from research laboratory to first commercial prototype. Today, fiber-optic amplifiers are considered basic building blocks available to all telecommunication network planners, who now intend to use them in a wide range of communications systems ranging from cable television networks to transoceanic telecommunications systems. The erbium-doped amplifier illustrates clearly that only by maintaining a broad scientific infrastructure (in this case in AMO science, laser engineering, and materials science) can a country expect to be a player in rapidly advancing technologies.

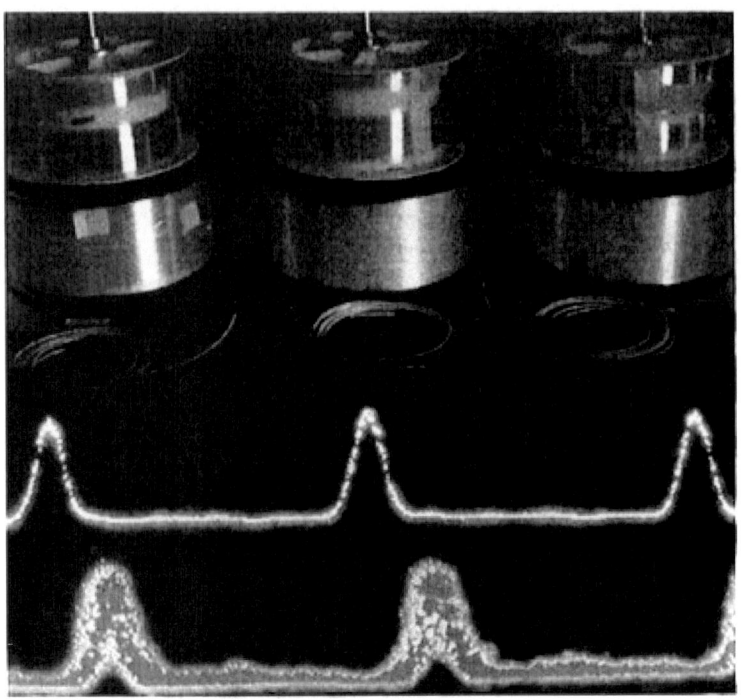

FIGURE 2.8 Erbium-doped fiber-optic amplifiers allow use of fibers to transmit bits of information in error-free form at unprecedented high rates for long distances. (Reprinted, by permission, from Peter Angelo Simon, "After Image," *Optics and Photonics News* 4 (January), 64, 1993. Copyright © 1993 by the Optical Society of America.)

There are a number of research opportunities in this area. For example, many optical communications systems operate with laser radiation having a wavelength of 1.3 μm, which is not in the 1.5-μm operating regime of the erbium-doped fiber amplifier, and AMO research aimed at developing optical amplifiers in the 1.3-μm regime could have great impact on this installed base of systems. Another area in which AMO science can have major impact is in the application of nonlinear optical techniques to improve optical transmission in fibers. Nonlinear soliton propagation has been shown to allow optical pulses to be transmitted over very long distances without any distortion.

Optical Data Storage

With the widespread consumer acceptance of CDs for high-fidelity stereo recordings, optical information storage is now familiar to nearly everyone. This storage technology is built on focused research on low-cost semiconductor lasers and the physics of the optical interaction of light with thin film materials. It is an important technology with many applications in the computer and printing industries. Although magnetic data storage dominates the data storage market today, it is anticipated that for many applications in the future magnetic storage will be completely replaced by optical disk storage. A pocket-size optical disk can hold 300 megabytes of data. They are removable and readily transportable and can simultaneously carry mass-replicated read-only information as well as user-rewritable information.

AMO science offers the potential for dramatic improvements in optical storage technology. Information storage densities can be improved by using shorter optical wavelengths in the optical recording and reading head because shorter wavelengths can be focused to a smaller spot size. Recent advances in blue-light-emitting semiconductor lasers, in highly efficient harmonic generation of light from semiconductor lasers, and in optical parametric oscillators hold the promise of providing the required radiation. Even the limitations imposed by the optical wavelength can be overcome. Normally, optical techniques cannot resolve structures that are smaller than an optical wavelength. Recently, however, the combination of nanoscale technology with optical physics has resulted in so-called near-field optical microscopy, which can resolve structures smaller than the optical wavelength, suggesting that new very high density optical storage systems might be realized in the future.

Optical logic devices and optical computing have been elusive objectives, but they provide an area in which real progress could have a large potential payoff. The combination of optical processing with electronic processing continues to represent an opportunity to exploit the best attributes of both. Progress in optoelectronic integration is especially important for lowering the cost of optical technology for mass deployment such as is needed for the practical realization of optical interconnect technology into computing systems.

> *AMO science plays a pivotal role in the development and characterization of devices and materials used in information technologies. Continued research will enable further technological improvements that will keep the U.S. information industry competitive in the global marketplace.*

Energy

Energy is a dominant national resource and a commanding international "currency." The availability of energy and its management for effective and safe use underlie the quality of life of people in all industrialized nations. The average person in the United States today enjoys energy benefits related to work, transportation, entertainment, home comforts, and conveniences well beyond those of any previous generation. To appreciate the role of energy in supporting a national position of influence and leadership, it is only necessary to recall the trauma of the "energy crisis" in the mid-seventies, to reflect on the current competition in international trade, or to recall the total necessity of various forms of energy to maintain a global military presence in a turbulent world.

Energy resources must evolve as known resources are depleted, as new technologies offer advantages of efficiency and environmental compatibility, and as expanding populations place ever-increasing demands on the quantity and quality of energy resources. This evolution of technology can proceed only by drawing on a fund of basic scientific and technical knowledge, much of which is based on the AMO sciences.

In this section, examples are presented of past and potential contributions of AMO science to the major energy technologies in use today (primarily combustion of fossil fuels supplemented by fission reactors and, to a lesser extent, by hydroelectric and solar energy), as well as those identified as possible energy sources of the future.

Energy Production

Combustion. Most of the energy used in the United States today is derived from organic and fossil fuels: oil, natural gas, coal, and renewables (e.g., wood). Taking account of the vast coal and oil shale resources of the nation and the technological feasibility of conversion to the currently used fuel forms, there is little doubt that organic fuels will continue to be a mainstay of the nation's energy diet for some time. Given the limited supply of fossil fuels, it is essential to learn how to use them as efficiently as possible. Because the burning of organic fuel accounts for a major part of the pollutant burden of Earth's atmosphere, it is equally necessary to learn to effectively control and limit the release of pollutants and to detect and monitor them. AMO science plays a vital role in achieving these goals.

The interiors of furnaces, combustion engines, incinerators, and other combustion reactors are hostile environments. It is difficult to put instruments into these environments without damage or destruction. But by means of laser diagnostics, it is possible to externally probe temperature, particle motion, molecular composition, and reaction rates in these hostile environments (Figure 2.9). These diagnostic methods are invaluable tools for optimizing the design and performance of such reactors.

The role of molecular physics in combustion research extends well beyond simple diagnostics. Spectroscopic investigations provide the detailed knowledge of energy levels and intramolecular processes needed to begin to understand the

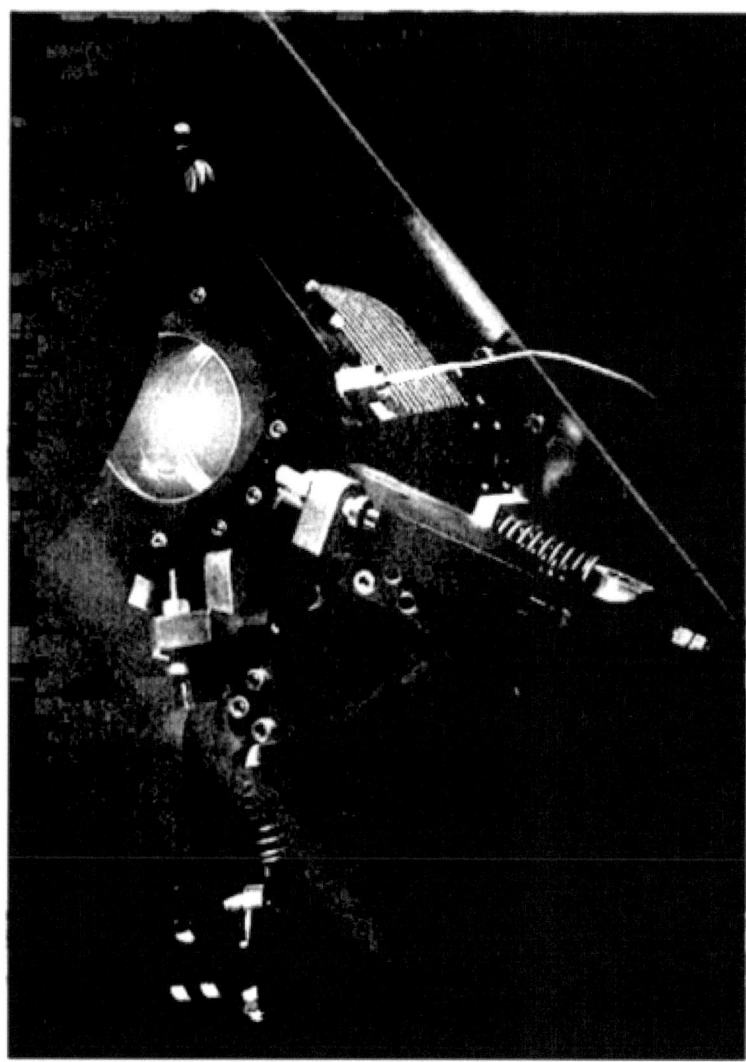

FIGURE 2.9 Detailed diagnostics of various forms of combustion chambers are extremely important to interpret the combustion process in terms of atomic and molecular processes occurring in the chamber. Atomic and molecular data are then needed to properly model the process and point to ways to optimize combustion. Here, lasers are used in diagnostics of an automobile engine. (Courtesy of Sandia National Laboratories, Albuquerque, N. Mex.)

course of chemical reactions. Experimental research in chemical kinetics yields quantitative information on the rates of the reactions that determine the course of combustion. Research in both kinetics and dynamics is contributing insight into these reactions at the molecular level. Elaborate models that often involve hundreds of reactive and inelastic processes make use of these data to simulate combustion. Although the modeling of chemical processes in simple laminar flames is now well established, a more ambitious goal remains. It is the development of truly predictive models of combustion that start from first principles, incorporate both molecular processes and fluid flow, and are capable of treating turbulent flames in practical configurations. The rapid increase in computational power suggests that now is the time to begin thinking seriously about what new molecular physics is required to support the development of these models. As an example, it will be necessary to know in detail the range of validity of statistical theories of chemical reaction rates, which entails understanding how quickly energy moves through molecules and whether the end result is random. This, in turn, requires breakthrough research in such areas as the use of femtosecond laser techniques to probe the evolution of highly vibrationally excited molecules. Many such examples exist; the potential future contributions of optical and molecular physics to optimizing combustion processes are great indeed.

Fission. The importance to the nation of maintaining a strong program in AMO science was illustrated during the energy crisis in the mid-seventies, when it appeared that there would be a shortfall in the nation's ability to keep up with the demand for reactor-grade uranium. AMO scientists were quickly mobilized, and two methods of laser isotope separation, one involving atoms and the other molecules, underwent intensive investigation. New techniques were devised for precision spectroscopy and for interpreting complex spectra, and laser systems were developed that could be precisely tuned to the required frequencies. The decrease in demand growth in the eighties converted the original objective to a more difficult one, namely, to enrich uranium at the world's lowest cost in order to maintain the U.S. share of the multibillion dollar enrichment market. The atomic vapor process was chosen for further engineering development at Lawrence Livermore National Laboratory. This program uses the world's most powerful visible laser system and has produced substantial quantities of reactor-grade uranium. This laser-based technology is on track toward commercial deployment as the existing gaseous diffusion plants are retired.

Solar Energy. The energy flux from the sun at Earth's surface amounts to about 1 kilowatt per square meter ($kW\ m^{-2}$). It is a generally clean and nonpolluting energy form and is attractive from many points of view. Many different approaches can be taken to harnessing solar energy, including direct conversion into electricity via solar cells, natural and artificial photosynthesis as occurs in plants, and solar heating and cooling of buildings. Solar energy also drives surface winds and ocean thermal currents that can be tapped as energy sources.

Even the proposed use of hydrogen for transportation fuel is in some measure solar if ocean thermal currents or other solar collectors are used to extract hydrogen from water.

The key to using solar energy for electrical power lies in being able to efficiently convert it to electricity and to store it for times when the sun is down. The technologies for both exist, and it is now a matter of making these economically competitive. The production of large-area amorphous silicon photocells by plasma deposition (see the section above on plasma processing) is one promising approach.

Photosynthesis is the way that nature converts solar energy into organic fuel. AMO scientists have been gaining insights into the interactions of light and matter and the molecular processes that occur in this conversion process. Some successful attempts at artificial photosynthesis have now been recorded, and both natural and artificial photosynthesis figure into the energy equation.

Fusion: Magnetic Confinement. Fusion, as it is envisioned, is a relatively benign energy technology in terms of its impact on the environment and threat to health and safety of people. The technology, however, is difficult and complex, and 40 years of research have just been rewarded with the demonstration of a "break-even burn," in which as much energy was produced through fusion as was used to create the burn. Thus, fusion is not yet a viable and economic source of energy. Further research in this area is merited, however, because the rewards of success will be great given an essentially limitless supply of fuel in the oceans.

Feasibility depends, among other things, on characterization and control of energy loss mechanisms in the reaction vessel, which can arise from interactions between electrons or photons and ions of heavy elements liberated from the reaction vessel by the various forms of radiation present. These energy loss mechanisms require new knowledge of the spectroscopy and excitation-recombination cross sections of atomic ions and information about the effects of strong electromagnetic fields on atomic properties and processes.

A valuable contribution of atomic theory to the design of magnetic fusion reactors has been the demonstration that the presence of small fractions of highly charged heavy ions can lead to significant energy loss. This has led to the replacement of heavy metals such as iron and molybdenum in the reactor walls, divertors, and delimiters by lighter materials such as graphite. Even with such low-z atoms, it is important to minimize wall erosion and the transport of such impurities into the central plasma region. Studies of plasma-edge effects have been recognized as being of paramount importance and are a significant component of the International Thermonuclear Experimental Reactor (ITER; Figure 2.10).

Improvements in the techniques of plasma spectroscopy have led to more detailed information on temporal and spatial variations in plasma properties such as electron density and temperature, the strength of magnetic fields, and the rate of flow of particles in or out of the plasma. Studies of the intensity and line profiles of the radiation emitted by the plasma provide detailed information concerning the impurity ions as well as dominant electrons and hydrogenic species present in the plasma core. Further data can be obtained by studies of the radiation induced by the injection of laser or particle beams. Studies of Thompson scattering of laser light and the photons emitted following electron capture by impurity ions from neutral hydrogen beams have been particularly valuable in this respect.

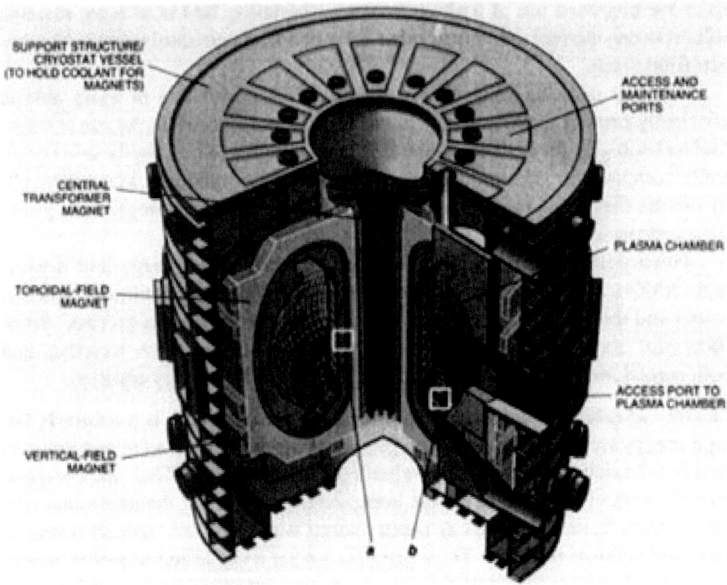

FIGURE 2.10 Schematic view of the International Thermonuclear Experimental Reactor. (Reprinted, by permission, from Robert W. Conn, Valery A. Chuyanov, Nobuyuki Inoue, and Donald R. Sweetman, "The International Thermonuclear Experimental Reactor," *Sci. Am.* **266** (April), 103-110, 1992. Copyright © 1992 by Scientific American, Incorporated. All rights reserved.)

Fusion: Inertial Confinement. The effort to achieve laboratory ignition of thermonuclear fuel by laser implosion of small pellets rests on a long history of AMO science advances that contributed to the understanding of dense plasmas and also led to the development and construction of the powerful laser sources needed to make the plasmas. In the case of lasers, these contributions range from the understanding of ionic spectra in solid-state laser materials and of excimer laser systems, to the linear, nonlinear, and electro-optical techniques developed for beam manipulation and control. X-ray and laser plasma diagnostics also play an important role in evaluating target performance. In addition, the contributions of atomic physics to the understanding of interactions in plasmas of highly stripped atoms—particularly for advanced targets employing complex mixtures of elements to control energy transport and hydrodynamics during the implosion—are important not only for laser-driven targets, but also for heavy ion drivers, which are viewed as attractive candidates for commercial energy production. Correlation of theoretical models with experimental data from existing laser target facilities, or from the next-generation laser driver proposed for reaching thermonuclear ignition, will be crucial to evaluating the feasibility of inertial confinement fusion for energy production as experiments push into regimes of higher plasma density than have been explored to date.

Efficient Use of Energy

Efficient use of energy is an extremely important part of any energy policy. Prior to the energy crisis of the mid-seventies, the nation had become extremely cavalier about energy use, and waste was rampant. Since that time, people have become more educated and caring about conservation, and this increased awareness has reaped rewards in lower energy consumption.

Conservation in lighting has been strongly affected by AMO science. The fluorescent lamp and arc lamp, widely used for lighting in businesses, homes, factories, and streets, function by the excitation and fluorescence of atoms and molecules in the gas phase, accompanied in some cases by conversion of ultraviolet light to visible at a phosphor on the walls. These lamps were initially developed without a detailed understanding of much of the physics involved, but now detailed gas-phase models and an understanding of gaseous electronics are used, together with observations, to improve and alter their properties. This progress toward a comprehensive understanding of lighting is a triumph of plasma physics combined with AMO collisional radiative theory and experiment. The biggest challenge for the future is to increase phosphor lifetime by reducing surface damage by particle bombardment, requiring further studies of particle-surface interactions.

> *Achieving harmony between energy production and use and a healthy environment is absolutely essential. However, the atmosphere is heavily burdened with carbon dioxide, a serious concern in terms of global change, and with the oxides of nitrogen and sulfur as a result of organic fuel consumption. With continued research and development, AMO science can reduce the environmental impact of energy production using fossil fuels*

through improved understanding of combustion reactions and combustion reactors and through better diagnostic procedures to detect and quantify pollutant emissions. AMO science is also central to the development of less environmentally harmful sources of energy such as solar and fusion power.

Global Change

Spiraling energy use and other human activities have led to measurable effects on the global environment. There is increasing international concern about the impact on Earth of human activities, in particular in the areas of global warming, depletion of the ozone layer, and pollution problems such as acid rain (Figure 2.11). As discussed in the section on atmospheric and environmental science, AMO science plays a pivotal role in the development of computer models to analyze the effects of changes in atmospheric composition and in the provision of reaction rates, collision cross sections, line shapes, and so on, for inclusion in such models. AMO science also contributes to reducing the load on the environment from energy use. Monitoring atmospheric constituents, modeling the impact of introducing foreign gases into the atmosphere, and predicting the eventual consequences of changes in atmospheric constituents are important challenges to AMO science. Global change considerations can also stimulate experimental and theoretical advances in AMO research. For instance, recent appreciation of the role of heterogeneous processes in the generation of acid rain and the destruction of stratospheric ozone has led to development of a number of novel experimental techniques to study mass accommodation and heterogeneous reaction processes on aqueous and aqueous-acid liquid and ice surfaces. The data obtained by these methods have led to a novel model of the mass accommodation of gaseous species onto liquid water surfaces.

The most difficult atmospheric challenge is to reliably quantify a wide variety of trace species, in both gaseous and condensed phases, which may be important to any given environmental problem. Most of the atmosphere is relatively nonreactive nitrogen, oxygen, and argon. The concern is the minor constituents of the atmosphere, that is, gaseous species that may constitute less than 1 part per trillion by volume (pptV) and condensed cloud droplets and aerosol particles that occupy a fractional volume of about 1 part in 10 million. Reliable quantification of important gaseous trace chemical species varies in difficulty. Carbon dioxide at about 355 parts per million by volume (ppmV) is relatively easy to measure, while the hydroxy radical at 0.5 to 0.005 pptV during daylight poses a much more difficult problem. Most other trace species of interest fall between these two extremes.

Until relatively recently, most atmospheric trace species were measured by grab sampling followed by batch analysis using either wet chemical or chromatographic techniques. However, advances in electro-optics (especially tunable lasers), microprocessor-based control electronics, and other technologies have recently spurred the development of highly sensitive, real-time spectroscopy-based measurement techniques for many key atmospheric trace species.

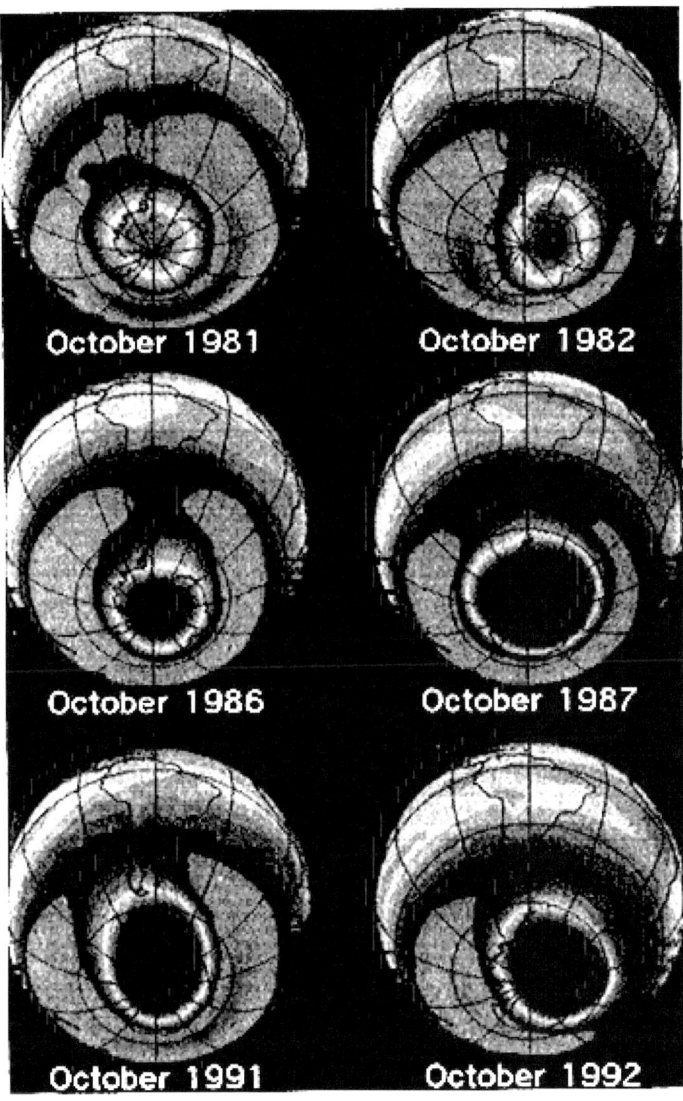

FIGURE 2.11 The ozone hole is expanding in size, as indicated here, with attendant risks that plants, animals, and humans will be exposed to excessive amounts of ultraviolet light. (Courtesy of National Aeronautics and Space Administration, Goddard Space Flight Center, Greenbelt, Md.)

The sensitive, real-time trace species measurements made possible by laser-based and other advanced electro-optical techniques do more than quantify the current chemical content of the atmosphere. Simultaneous measurement of interrelated species with good time resolution allows detailed tests of the photochemical mechanisms embedded in atmospheric models. Furthermore, coupled with micrometeorological measurements they allow direct determination of key trace gas fluxes between the atmosphere and Earth's surface, permitting experimental confirmation of critical atmospheric model boundary conditions.

Atmospheric laser-radar (LIDAR), a form of remote laser spectroscopy, provides a graphic example of the use of AMO science in field measurements and is the oldest field application of laser spectroscopy in atmospheric remote sensing. In its simplest form, fluorescence LIDAR, a laser is tuned to a resonance excitation of the species of interest. The laser is projected into space, and the fluorescence from excited species is collected and detected by a telescope at the laser source. By collecting the signal as a function of time, the density as a function of distance can be measured. For example, even in the early seventies, investigators were detecting sodium atoms in the mesospheric layer 80 to 100 km above Earth's surface. This region is too high for balloons and too low for satellites, and previously could be accessed only by rockets.

LIDAR provides the opportunity to study conditions on a continuous basis. Mesospheric LIDAR has profited immensely by the development of stable, precise laser sources. By using lasers that are locked in absolute frequency with megahertz accuracy, observers are able to remotely measure not only densities but also wind velocities with an accuracy of 3 m sec^{-1}. In addition, by accurately recording the profile of the excitation lines and deducing the population of hyperfine levels of the ground state, the temperature of the mesosphere can be measured.

The mesosphere is an area of fundamental importance to atmospheric physics. The measurement of winds and densities in the layer allows observation of atmospheric gravity waves reflecting the dynamics of the atmosphere. In addition, mesospheric measurements are of importance to both global warming and ozone depletion studies. Carbon dioxide and methane, which lead to warming of the troposphere (the greenhouse effect), lead at the same time to cooling of the top of the atmosphere including the mesosphere. It is expected that over the next century, while the troposphere warms a few degrees, the mesosphere will cool by as much as 20° Celsius. Already there have been reports of a steady lowering of the mesospheric boundary consistent with a cooling of that layer. This may be an early measure of a basic change in the atmosphere.

The sodium (Na) atoms in the mesosphere come primarily from meteoric dust. The sodium atoms react to form NaO, NaO_2, and NaOH, which migrate into lower regions of the atmosphere. There, reactions with hydrogen chloride (HCl) result in the release of chlorine (Cl) as a free atom, which them enters the catalytic ozone destruction cycle. It is now thought that mesospheric sodium may be a significant contributor to the complex balance of processes that determines the rate of destruction of ozone in the upper atmosphere.

LIDAR techniques have also been used in many other remote sensing applications, particularly for monitoring pollution sources. For instance, LIDAR has been used to monitor the emission of sulfur dioxide and other gases from power plant smokestacks; it allows off-site monitoring and precise pinpointing of sources. Law enforcement officials have expressed an interest in the use of LIDAR, for example, to detect effluents characteristically associated with clandestine drug operations, and there is interest in extending the techniques to enforce nuclear nonproliferation, as well as to detect biological and chemical warfare violations. An area of major potential impact is the use of remote sensing devices for monitoring the evolution of gases in nuclear waste storage depots, where direct access is limited. In all these areas, LIDAR promises to serve as one of a family of tools brought to bear on problems of immense concern to the nation.

Remote sensing of atmospheric constituents and atmospheric pollutants using LIDAR and other techniques is a critical part of the global change research effort. Advanced field systems for ground-based monitoring and advanced in situ detection systems for finding trace and unstable species are urgently needed and are dependent on the availability of stable, tunable, field-qualified laser sources of sufficient power and energy output. Problems in local and regional air pollution, lake, river, and marine pollution, groundwater contamination, and toxic waste disposal all pose similar modeling, field measurement, and laboratory and theoretical challenges, and all benefit from AMO science. Even ecological problems with a large biospheric component and environmental health issues with a major focus on human physiology still require sophisticated chemical and physical measurements. Determining the relevant interactions of atoms, molecules, aerosols, and sunlight is a great challenge to AMO science, as is the task of developing computer models that will predict the effects of pollutants.

Defense

AMO science has been, and will undoubtedly continue to be, important in national defense. Lasers and electro-optics have come to the forefront in a variety of ways in modern warfare. In addition, atomic and molecular processes and spectroscopy play an important role in such areas as radio propagation in the ionosphere, atmospheric infrared transmission, infrared background emissions, radiation trapping in nuclear bursts, nuclear blackout of electromagnetic propagation, and missile plume detection and analysis.

Significant among the early applications of lasers were rangefinders and target designators; these applications were based on the ability of the laser to produce short pulses of radiation and to be pointed precisely at distant targets. These early applications have expanded to include laser systems for optical countermeasures, gyroscopes, range-resolved Doppler imaging radars, illuminators, space communication systems, fiber-guided missiles, and fiber sensors. In the area of C^3 (command, communication, and control), the Department of Defense (DOD) has both contributed to and benefited considerably from fiber-optic communications technology. In recent years, research programs have been under way to apply lasers to optical interconnects and optical information processing. It is noteworthy that, in all of these areas, a good deal of the laser technology and applications development that has been carried out with DOD funding has proved useful in nonmilitary areas.

Weapons Systems and Delivery

Laser rangefinders are used to measure distances for ground-based munitions; Q-switched Nd: YAG solid-state lasers producing pulses of the order of 5 to 10 nanoseconds are capable of range resolution of the order of 1 meter. The new generation of rangefinders will use semiconductor laser diodes or diode-based eye-safe solid-state lasers.

Target designators employ laser beams reflected from a distant target to guide a bomb or missile containing a smart seeker onto the target. These systems have proved important in the precision guidance of munitions. They were used successfully in the Vietnam War and in Desert Storm to avoid collateral damage. Lasers have also been used to illuminate distant targets for optical surveillance and reconnaissance.

Atomic opacity is an important problem in nuclear weapons physics. There have been few experimental measurements of the opacity, that is, the degree of absorption of transmitted light, for plasmas at moderate or high temperatures and density. The calculation of opacities requires a large atomic or molecular database, with information on the energy levels of several ionic stages of each species in the plasma and of the corresponding radiative transition probabilities and line shapes. Available results are not of sufficient accuracy for current needs, especially for many-electron atoms, and do not take advantage of the large increases in computer power and the advances in theoretical techniques achieved in recent years.

Two efforts are under way to provide better opacity data. One project, carried out as a collaboration between the United Kingdom and the United States, exploits the configuration-interaction codes for atomic structure and matrix codes for continuum wave functions. Although this effort has so far produced few new opacity data, the physics underlying these codes is excellent, and valuable intermediate results have been obtained, such as collision cross sections, energy levels, and transition probabilities. Another approach has been the calculation of atomic wave functions using a model potential. The first results of this technique have resolved several outstanding problems in astrophysics, and the method is being evaluated currently for the high-temperature plasmas appropriate to fusion and defense applications.

Remote Sensing

AMO instrumentation is crucial to several defense programs for remote sensing. This includes laser-radar (LIDAR) monitors for chemical and biological warfare agents and fiber-optic sensors. Pulsed Doppler laser radars using heterodyne detection have been successful in providing complex, time-resolved images of distant scenes in the six-dimensional space of position and motion. The utility of laser radars for terrain-following guidance and obstacle avoidance for low-flying aircraft has been demonstrated. Replacing microwave radars, these LIDARs allow aircraft to avoid detection.

Lasers tuned to specific infrared wavelengths provide effective monitoring systems for chemical and biological warfare. Fear of the use of such warfare techniques on embattled populations can be reduced with effective remote monitoring techniques, using the concepts of AMO science. Further discussion of this topic is found in the section on global change. Fiber-optic sensors have proved particularly effective as acoustic sensors, providing much more accurate undersea measurements than traditional sonars. Fiber sensors can remotely and/or passively detect motion, strain, temperature changes, and magnetic fields and can be coupled to optical detectors. These versatile sensors can be used in many places, from the battlefield to embedded in the skins of aircraft.

Atmospheric transparency is important in the design and operation of infrared systems such as the forward-looking infrared system (FLIRS), heat-seeking missiles, and infrared lasers. Data on atomic and molecular processes are vital to the understanding of atmospheric and meteorological phenomena that affect military scenarios. Much effort has gone into developing both highly resolved (HITRAN) and broadband (LOWTRAN) transmittance models. This has involved the measurement and calculation of line positions, lower-level energies, line strengths, and pressure broadening and shifting coefficients for water vapor, carbon dioxide, and other absorbing molecules occurring in significant quantity in the atmosphere.

Infrared signatures of missiles are being examined through spectroscopic analysis of rocket motor plumes including measurements of rotational temperature and features such as the blue spike in the 4.5-mm carbon dioxide band. In recent years, these techniques have been extended to monitoring the emissions accompanying a missile launch, fuel spill, ground test, and manufacturing process. In the case of a missile launch, the monitoring activity is combined with mobile tracking and computer modeling to predict the shape, size, and concentration of gases in a plume as a function of local conditions such as meteorology and terrain. One goal is to determine where or when certain operations should not be carried out because of possible exposure hazards to local populations.

Countermeasures

Countermeasures on the electro-optical battlefield are required to defeat commonly used infrared and near-infrared sensor and imaging systems. Effective laser countermeasures to such optical systems are currently under extensive development. Defeating visible and near-infrared imaging systems and missile heat seekers is the major thrust of these efforts. Because of the intensity gain due to focusing and imaging in optical systems, countermeasure systems can use relatively low power lasers. Infrared heat-seeking missiles (SAMs) represent a serious threat to both civilian and military aircraft that operate in or near areas of strife. When used against civil aviation, they can become a weapon of terror. Currently, military aircraft are equipped with missile countermeasure systems that are successful against early generations of heat seekers. However, newer heat-seeker systems have been designed to counter these countermeasures. It is a challenge to the countermeasure designer to develop new systems that do not have the shortcomings of those is current use. A laser can be used either to permanently damage or to temporarily dazzle sensors, rendering them ineffective. This technique can be used to take out of commission visible and near-infrared systems, which are used in surveillance, reconnaissance, discrimination, weapons direction, and guidance. Typical sensors that can be blinded are the human eye, charged-coupled device (CCD) focal plane arrays, and image intensifiers that require lasers in the 0.4- to 1.0-mm range.

C3—Communication, Command, and Control

Fiber optics are being used increasingly in military communication, command, and control systems because of their high bandwidth, small size, freedom from noise and cross talk, ability to withstand electromagnetic impulses, and the security provided by their low radiant emission. They interconnect computer systems within command centers, within aircraft and within ships. With the increasing ability of computers to handle the vast amounts of information that electro-optical systems make available, the high bandwidth of laser-driven fiber-optic communications becomes crucial. This makes possible increased "fusion" of the data from many sensors into intelligible, readily used signals to improve battlefield response. Because fibers are so small and flexible, they can be used effectively to guide missiles, allowing them to be remotely maneuvered in real time; such systems have been deployed and are in the military inventory.

The use of fiber-optic communications for interconnecting computers is well established. Interconnects within computers on a subsystem level are under development, and board-to-board and chip-to-chip optical communications are active research areas. These interconnects will move down into smaller subsystem levels as more efficient laser sources are developed. Microcavity lasers will perhaps be used here. Interconnect research is driven by the need for teraflop (1012 floating point operations per second)-level computation power for automatic target recognition, antisubmarine warfare, electronic warfare, and electronic intelligence, together with the fact that interconnect bottlenecks limit the development of teraflop systems using silicon technology. The high degree of parallelism, as well as the speed of optics, also points to the potential for the direct use of optics in processing.

The infrared, visible, ultraviolet, and X-ray background of the atmosphere and its variability with time and location must be accurately delineated to help in modeling the operation of sensors and communication systems. Here also, a vast collection of spectroscopic data is needed, and considerable effort is required to understand the effects that high-altitude atomic and molecular dynamics have on this background.

Disturbances in the atmosphere due to nuclear explosions can cause major changes in its ionization and thermal and chemical structure. The increase in ionization leads to absorption of radio waves, and a radio blackout occurs, in which communications are disrupted. The disturbed atmosphere is a copious emitter of radiation, which can serve as a remote diagnostic probe of the changing environment. The atmosphere, as it responds, can be regarded as a dynamically evolving laboratory of AMO science. The description of its evolution and its return to the undisturbed state depends on a complex variety of atomic and molecular processes involving ions, neutral atoms, and charged and neutral molecules in abnormal distributions of energy levels. With the atmospheric test ban and the moratorium on underground testing, laboratory simulations of the highly disturbed atmosphere are necessary.

Navigation systems for the military increasingly use technologies having AMO science as their base, including gyroscopes and atomic clocks. The laser gyroscope has both civil and military applications in guidance and navigation. Active and passive laser gyros have been shown to be useful in applications that require high rate and/or high stability; being nonmechanical they are not subject to wear and tear. All-solid-state fiber gyros are expected to provide inexpensive replacements for mechanical gyros in a number of guidance applications.

The cesium atomic clock is used on the Global Positioning System, which provides information on positioning that is vital for military commanders. This technology is rapidly becoming available to consumers. Future ideas for atomic clocks include solid-state silicon-based microcavities. These clocks need to be reliable and inexpensive.

National defense and security are dependent on contributions from AMO science and will continue to be as new optical technologies are introduced. Further, simulation of warfare scenarios is becoming increasingly important in a downsized military establishment, and there is a clear need for a wide variety of accurate atomic and molecular data that describe normal and disturbed atmospheres and for variability models to simulate sensors and other electro-optical systems.

Health and Medical Technology

Medicine

Lasers. Lasers have become indispensable tools in numerous therapeutic and diagnostic medical procedures. A laser beam can be precisely focused at discrete lesions in the eye, the skin, or other exposed areas and can also be directed through thin flexible optical fibers, allowing, in principle, laser treatment of any body organ through a minimally invasive procedure (Figure 2.12). Today lasers

FIGURE 2.12 Micromanipulators used to deliver laser beams of very small diameter for precision work under the operating microscope. These micromanipulators accept the output of the laser and project a visible aiming beam, as well as the treatment beam, onto the surgical site in the field of view of the microscope. Controls on the micromanipulators allow the surgeon to move the beam within the microscope's field of view and to change the spot size. (Courtesy of Coherent, Inc., Palo Alto, Calif.)

are used to treat a wide variety of medical problems and frequently provide an attractive alternative to surgical intervention. With further research into treatment methods and development of low-cost laser systems tailored to medical needs, it is clear that lasers will become even more important in medicine in the future.

The use of lasers in medicine and surgery has expanded rapidly for two reasons. First, a broad range of lasers have become available, allowing matching of the wavelength and temporal characteristics of the radiation to each particular clinical problem. Second, there has been considerable progress in understanding laser-tissue interactions. Laser energy deposition results in localized heating that can, for example, cause blood vessels to coagulate or lead to tissue ablation. At very high pulsed laser intensities, optical breakdown occurs, resulting in plasma formation and almost complete absorption of the laser pulse in a tiny volume.

Research is under way using low-intensity laser radiation to detect and treat cancer. Tumors tend to collect and store certain body pigments such as porphyrins, and cancerous tumors will take up and retain exogenously administered pigments and certain fluorescent dyes. Thus an early-stage cancerous mass can be detected through its concentrated fluorescence under blue or near-ultraviolet laser irradiation during, say, a fiber-optic catheter examination. A similar procedure can also be used to selectively kill cancerous cells. Illumination of porphyrin molecules with red light can result in the production of singlet oxygen, which is highly toxic. Thus, porphyrin-laden cancer cells can be destroyed by irradiation with red laser light, whereas interspaced normal cells are essentially unaffected. Another laser application in cancer therapy is laserthermia. Cancerous tumors are poorly supplied by blood vessels compared to normal tissues and cannot readily disperse heat. Thus, it is possible to locally heat and destroy a tumor using laser radiation introduced by an optical fiber.

Lasers are finding numerous applications in ophthalmology. Photocoagulation using an argon ion laser is now the standard treatment for diabetic retinopathy, although diode lasers are also now being considered for this application. Various techniques involving small laser perforations are being used to treat glaucoma. A common problem following cataract surgery is that the membrane holding the implanted intraocular lens becomes opaque. Optical breakdown induced by a Q-switched Nd:YAG laser is used to cut the membrane, removing it from the optical path. Studies of the cutting process have revealed a complicated series of events initiated by laser-induced breakdown, which can cause cavitation, shock waves, and liquid jets and create unwanted damage. These studies have led to an understanding of the scaling laws for such damage and have prompted investigation of the use of less energetic picosecond pulses for initiating optical breakdown. Picosecond pulse microsurgery has application in cutting unwanted structures near the retina, which would be damaged if nanosecond pulses were used. Lasers are also used in correcting vision defects, most of which are due to small anomalies in corneal curvature (the curved corneal-air interface provides ~70% of the total refractive power of the eye). Research is under way using argon fluoride (ArF) excimer lasers to reshape imperfect corneas through selective ablation. In addition to reducing the need for eyeglasses, this technique can also be used to remove corneal scars.

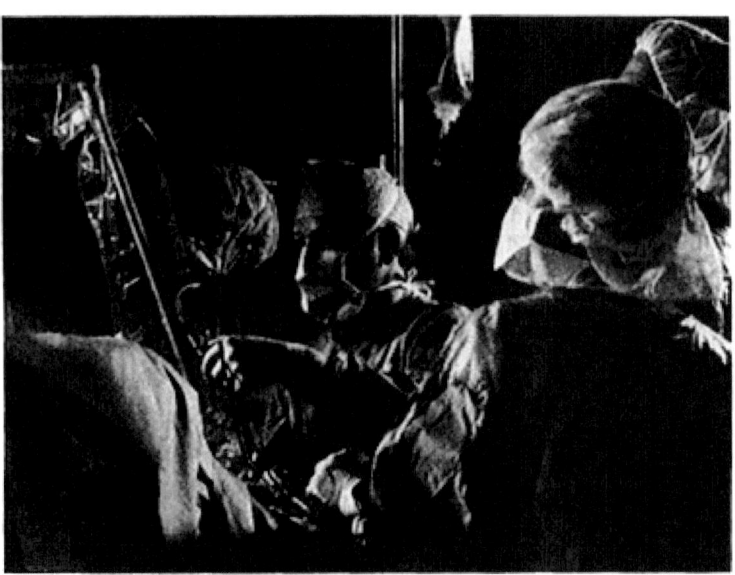

FIGURE 2.13 CO_2 laser being used to treat endometriosis. The operating laparoscope delivers the laser radiation directly to the tissue. The laser radiation serves as a scalpel, cutting adhesions and endometrial tissue with minimum damage to surrounding healthy tissue. (Courtesy of Coherent, Inc., Palo Alto, Calif.)

Lasers are also used as a scalpel (Figure 2.13). If the laser parameters and focusing are chosen such that the beam is intense enough to ablate and cut through the material directly in its path, and provide sufficient heating of neighboring tissue to cause coagulation, incisions can be made with minimal bleeding, which is especially valuable for surgery on vascular organs. If a short-pulse (`1 msec) laser is employed that is strongly absorbed in the target, the absorbing volume ablates before there is time for appreciable outflow of heat, and consequently each pulse precisely removes a thin layer of material. For example, lasers have had a dramatic positive impact on the treatment of gynecological diseases in women. The precise application of intense laser energy permits destruction of only damaged tissue while maximizing preservation of the normal reproductive tract. A number of new laser technologies are finding their way into clinical use. The pulsed Er: YAG laser (with a wavelength of 2.9 µm) has been shown to precisely cut both bone and soft tissue. The pulsed Ho: YAG laser (with a wavelength of 2.1 µm) is not as strongly absorbed by tissue water, resulting in greater residual tissue damage, but has the advantage that it can be transmitted by conventional quartz fibers, allowing its use in procedures requiring fiber-optic delivery of laser light. Thermal and acoustical effects of pulsed lasers are being actively modeled, and deleterious effects, such as the generation of gas bubbles, are being studied. Diode laser systems are also now reaching output power levels suitable for use in tissue cutting and welding.

Both Ho:YAG and xenon chloride (XeCl) lasers have been used in clinical trials as an alternative to balloon angioplasty. In laser angioplasty, pulsed-laser ablation is used to clear obstructing arterial plaque. However, this requires precise removal of as much diseased tissue as possible without puncturing the vessel wall. Diagnostic feedback methods such a ultrasonography and tissue spectroscopy are being studied for this role. The acoustic shock waves that result from pulsed-laser ablation are used to clinical advantage in the treatment of kidney stones. Stones lodged in the urinary tract are often fragmented by shock waves when illuminated with visible pulsed dye laser light. Such laser procedures provide an attractive alternative to surgery.

Medical Imaging. Medical imaging frequently employs X-rays with the attendant risk of cancer induction. As an alternative, researchers are exploring the possibility of imaging using light. This is a difficult problem because strong multiple scattering occurs in tissue, destroying the image contrast and making it difficult to extract spatially resolved information from measurements of the transmitted or backscattered radiation. However, multiply scattered photons travel a greater distance in the target before emerging than do those transmitted directly, and recent work has demonstrated that illumination with picosecond laser pulses, in combination with the use of an ultrafast optical gate, provides a means to selectively discriminate against multiply scattered light. Numerous variations of this simple idea are currently being explored. Time-gating has been used for optical transillumination of tissue, and the images obtained have been investigated for applications such as the early detection of excess blood in the brain (detection of strokes) using multiple source-detector pairs to locate the source of absorption by blood. Knowledge of the absorption spectrum of blood can be used to guide the selection of wavelengths optimal for detection of oxyhemoglobin or deoxyhemoglobin. Picosecond lasers are also being used in the early diagnosis of breast cancer. Initial imaging experiments used mode-locked dye lasers as sources and streak cameras or fast microchannel devices as detectors. Such time-domain imaging involves expensive lasers and detectors and is quite complex. As a consequence, a number of laboratories are now investigating the use of simpler frequency-domain techniques in imaging.

Femtosecond laser pulses have been used to obtain optical echoes from structures within the eye and from different layers in skin. Such optical ranging has demonstrated better spatial resolution than is currently available using ultrasound techniques; however, the need for ultrafast pulsed lasers and gated detection limits the clinical utility of the technique. Recent work has shown that ranging measurements can be made by using a continuous wave low-coherence-length superluminsescent diode in an interferometer system. Measurements have been made on eyes and arteries in vitro, and a number of ophthalmic applications, such as early detection of damage to the optic nerve due to glaucoma, are being investigated. Femtosecond pulse technology is also being used to study the rapid processes involved in vision.

Cell Manipulation. Optical trapping (see Chapter 1) has a myriad of potential research and therapeutic uses in medicine, and commercial optical trapping systems have recently become available. Specifically, optical traps are now being used to study the forces involved in the locomotion of biological macromolecules and to manipulate and position cells ("optical tweezers"). Cells can be held for fusion or perforation with a second laser to facilitate genetic manipulation or in vitro fertilization.

Monitors. Measurements involving isotopic tracers are widely used in both clinical medicine and basic research. For example, this approach is being used to monitor the metabolism of calcium in the body, a process of interest in many areas, ranging from the study of osteoporosis to the feasibility of long-term spaceflight in zero gravity. Isotopically labeled calcium is consumed in the form of milk or other foods and subsequently measured in bones or blood. Traditional techniques that make use of radioactive tracers are clearly not desirable, especially in studies involving growing children and pregnant women. A better approach is to use a stable calcium isotope, such as ^{48}Ca. The ^{48}Ca atoms present in a blood or bone sample can be monitored with high sensitivity using high-resolution mass spectroscopy or laser spectroscopy. (These spectroscopic techniques can also be used to monitor heavy metals and other species present in the blood.) Studies using isotopic tracers require the availability of large amounts of isotopically enriched material, but these can be obtained by using laser isotope separation.

Radiation and Health Physics

Health physics deals with understanding the interaction of ionizing radiation such as alpha particles, beta rays, gamma rays, and neutrons with living systems and with the design of instrumentation to measure sources of radiation and estimate possible harmful effects on living systems. Such studies rely heavily on AMO science and are vital to human society because ionizing radiation occurs in the environment in the form of natural radiation and cosmic rays and also because it is introduced by human activities such as medical and industrial uses of radiation and nuclear energy technology.

Once energy is deposited in living tissue by ionizing radiation, a complex sequence of events occurs starting at the atomic level, continuing through a chemical phase, and ending with the observed biological and medical effects. Microscopically, collisions of energetic particles with atoms and molecules result in the production of excited atoms, ions, and secondary particles, most importantly electrons. Analyzing the subsequent chain of events and their possible effects requires a detailed spectroscopic knowledge of the pertinent excited and ionized states and the cross sections for all the major collision processes operative, remembering that the predominant chemical reactions in ground and excited-state collisions may be different. Further research is required to establish a more complete database of cross sections for collisions of electrons, protons, and heavy ions with molecules, especially polyatomic molecules of biological importance such as water and the hydrocarbons.

AMO science is also essential in the design of dosimeters to measure radiation fields and sources. Because many such instruments are dependent on excitation or ionization of some medium, research on the interaction of radiation with matter is required to calibrate and interpret the readings of such instruments. One outgrowth of dosimeter research was the development of resonance ionization spectroscopy, which can provide isotopically selective trace element detection and is now used to monitor accidental emission of long-lived radio isotopes and other species.

Design of Bioactive Molecules (Pharmaceuticals)

Molecular theory is making numerous contributions to the design of bioactive molecules, including drugs for treatment of disease, and herbicides and pesticides for agricultural use. The principal "tools" of molecular theory encompass quantum chemical techniques and a variety of methods that make use of empirical potential-energy functions, such as molecular dynamics and Monte Carlo simulations. The goal of most applications of these methods in the design of bioactive molecules is a determination of the geometric and electronic structural properties of the molecules of interest. Two overriding concerns in this regard are the electronic characteristics (e.g., charge distribution, dipole and quadrupole moment, and molecular electrostatic potential) and the nature of the bioactive conformation(s) of a molecule. Quantum chemical methods are playing an important role in dealing with these concerns for a wide variety of small molecules of biological interest. Due, however, to the significant number of torsional degrees of freedom possessed by a substantial number of these molecules, it is necessary in many cases to employ a "hybrid" procedure, which uses empirical potential-energy-function-based conformational searches followed by some form of quantum chemical calculation to determine the required electronic properties of appropriate conformers. In cases in which suitable potential-energy functions are unavailable, semiempirical quantum chemical methods have been used, but, as noted above, these methods are restricted in the size and conformational complexity of the molecules that can be treated.

The development of transferable potential-energy functions, coupled with the rapid growth in computational power of today's high-performance computers, has extended the range of problems that can be treated by molecular theory methods to include bio-macromolecular systems such as proteins and nucleic acids. Of especial interest here are a number of molecular dynamics and Monte Carlo-based techniques (e.g., thermodynamic integration and perturbation methods) for calculating thermodynamic properties such as the free energy of ligand-protein binding—a property that is notoriously difficult to calculate. These techniques have also been employed in studies of, for example, the solvation and conformational free energies of small molecules in aqueous and other solvents.

Advances in ab initio electronic structure theory in combination with new computing capabilities now permit approximate but realistic calculations on sizable molecules. In particular, such calculations can guide the synthesis of new drugs and can suggest strategies for creating new bioactive molecules. However, improved quantum mechanical methods for treating larger systems by semiempirical or ab initio techniques are required, together with development of improved parallel molecular dynamics algorithms to allow for more realistic simulations and for treatment of supramolecular systems such as biological membranes.

AMO science is important in a number of health-related areas, especially medicine. With the development of new low-cost lasers and delivery systems, laser-based procedures will become more widely accessible in the future. Also, new optical techniques promise to reduce the reliance on X-rays in medical imaging. Molecular theory is making significant contributions to drug design.

Space Technology

The U.S. space program has sizable efforts in astrophysics, space science, and atmospheric and environmental science as well as in a number of other areas, and the impact of AMO science in these areas has already been emphasized. In addition, AMO science, particularly through lasers, plays a strong role in space technology.

Measurement and Sensing

Until a few years ago, virtually all sensing from space was done passively. Active remote sensing systems using space-based lasers (LIDAR) are expected to play an increasingly important role, providing information for meteorology and pollution monitoring as well as long-term studies of global climate change.

LIDAR is a form of optical probing by light scattering that can be used to measure the total aerosol distribution and the concentration profiles of aerosols and trace gases, as well as wind vectors in the atmosphere. For the past several years, NASA has been developing the Laser Atmospheric Wind Sounder (LAWS) LIDAR system for the worldwide mapping of winds from space. Another program, LIDAR In-space Technology Experiment (LITE), is aimed at range-resolved studies of cloud top heights, a parameter important in weather forecasting. LITE II, a follow-on system, is being designed using a tunable Ti:sapphire laser for vertical profiling of the atmospheric temperature and water vapor content.

Lasers are used to accurately measure and control baselines in coherent interferometry, to obtain high angular resolution in the microwave region. These uses may be for astronomical applications or for tracking satellites at great distance. Research is under way to connect several microwave antennas with lasers by means of fiber-optic systems, but achieving this goal requires a detailed understanding of the noise properties of semiconductor lasers modulated by ultrastable microwave sources and the effects of fiber-optic transmission.

Spacecraft Navigation and Communication

Satellite-based laser communication links are being explored because they can be made very directional. This capability offers security, of considerable military significance, and the potential for very long distance communications, as needed for the deep-space network. Laser systems are projected to require lower power, weight, and volume than comparable microwave links. The military plans to fly diode-pumped Nd:YAG lasers in the first generation satellite-to-satellite communications technology, but these should ultimately be superseded by semiconductor diode lasers. The challenge in satellite communications is to achieve ultrareliability over long periods not only in laser performance but also in pointing and tracking. The resolution of these technological issues promises spinoffs into commercial applications. For example, diode-pumped solid-state lasers can provide ultrareliable sources with a wide array of uses in manufacturing and medicine.

The introduction of deep-space satellite missions using radio-frequency signals for communication was the impetus for a field known as radio science, in which the radio communication signals are used to extract scientific information about the intervening medium. The development of communication systems using optical signals enables an entirely analogous field of "light science." This field includes the study of planetary atmospheric absorption at optical wavelengths, fine-scale scattering from planetary ring systems, and integrated forward scattering over interplanetary distances in the solar system dust field. Furthermore, many previous radio-frequency experiments that were contaminated by charged particle density fluctuations from the solar wind, like gravitational body bending of electromagnetic waves, can now be performed without those disturbances, because the effects of charged particle fluctuations fall off as the square of the carrier frequency.

The development of solar-pumped lasers is being pursued for space power transmission and propulsion. New potential applications of solar lasers in space are emerging. These include earth, ocean, and atmospheric sensing from space; detecting, illuminating, and tracking hard targets in space; and deep-space communications. Gas, liquid, and solid lasers have all been considered as candidate solar lasers, and successful lasing has been achieved with the use of a number of such systems. It is predicted that solid-state lasing materials with broadband absorption characteristics such as alexandrite or Nd:Cr:GSGG might yield greater than 10% conversion efficiency.

High-performance satellites need massive on-board information-processing capability. Fiber-optic communication systems are expected to take their place naturally in the computers of spacecraft, as they do in ground-based systems. Optical information processing for on-board analysis and synthesis of sensor data is becoming increasingly important in satellites. NASA has programs in optical implementations of on-board signal processing, image processing for robot vision systems, and spatial light modulators as devices needed to achieve these objectives, as well as in fiber-optic communications. Long-range plans for lunar colonization and Mars exploration point up the need for these advanced optical technologies.

Optical sensors are needed in a variety of satellite applications. The fiber-optic rotation sensor, particularly in its integrated optics configuration, will be important in high-sensitivity, low-weight inertial navigation systems. Optical systems are being developed to assist in docking between satellites. Fiber optics and integrated optics sensors are also being considered for use in monitoring the health of a spacecraft.

Many spacecraft require accurate on-board clocks. For example, the cesium atomic clock is used on the Global Positioning System satellites. Recent developments in the field of atom-ion trapping and cooling suggest that with further research greatly improved spacecraft clocks can be developed that will offer much higher accuracies.

The space shuttle glows in the dark, and so do orbiting satellites. The glow is accompanied by erosion of the surfaces of the spacecraft and the instruments it carries. The glow complicates the design and operation of optical and infrared instruments, and the erosion limits their effective lifetime. The glow and erosion are produced by the impact of atmospheric ions and atoms with the spacecraft at velocities that correspond to energies of about 5 eV. This phenomenon results in a complex array of surface and gas-phase reactions that depend on the nature of the surface. Much further AMO research is, however, required in order to fully understand the phenomenon and minimize its effects.

One increasing problem in the space program is the growing amount of debris in orbit. Lasers have been suggested both as optical monitors of space debris (through a form of laser radar) and as a means for elimination of small amounts of debris (either by vaporizing it or by deflecting it out of the satellite path). In either case, lasers must be placed in the satellites. Further research and development are required to determine the optimal design for such applications.

AMO science plays an important role in spacecraft operations and applications that will increase in the future with the advent of sophisticated atmospheric remote sensing systems and laser satellite communications.

Transportation

The pace of our nation's commerce continues to rely heavily on an aging, yet critically important, transportation infrastructure. Unlike other segments of our national economy, the responsibility for this infrastructure rests heavily with federal, state, and local governments. At a time of increasing demands on scarce governmental resources, it is crucial that cost-effective technologies be available to improve the safety and effectiveness of transportation systems throughout the nation.

Although AMO research is not usually considered central to transportation issues, it nonetheless is playing an important role in air, land, and sea transportation through sensing and control of vehicle movement (navigation), through improvements in safety, and through increased fuel efficiency.

Aviation

In aviation, lasers are being used to detect wind shear, clear air turbulence (CAT), and wake vortices. These phenomena can cause severe injury to passengers and damage to aircraft. For example, the 1987 crash of a Delta airlines jet at the Dallas–Fort Worth Airport was attributed to severe wind shear. High-sensitivity light scattering techniques developed by AMO researchers are being applied to the detection of wind shear. For instance, a LIDAR system can measure wind velocity through minute frequency shifts of laser light scattered from moving aerosol particles in the atmosphere. Wind shear velocities are in the range of 10 to 30 m sec^{-1} and give frequency shifts of 1 to 3 megahertz (MHz) for 10-μm carbon dioxide lasers and 5 to 15 MHz for 2-μm solid-state laser sources. LIDAR systems can determine wind shear and turbulence out to a range of 10 km in front of aircraft and give pilots sufficient warning to take appropriate avoidance action.

Measurement of air turbulence has implications for the efficiency of use of airport resources. Airport operators need information on air turbulence coming from wake vortices created during takeoff and landing of large transport aircraft.

Because of the danger from this turbulence, aircraft must be spaced sufficiently far apart that the wake of the preceding aircraft does not disturb the next flight; such spacing can produce a severe bottleneck at congested airports.

Likewise, collision avoidance through automated systems, or through a warning system activated by laser radars, would be a considerable boon in skies near airports. Lasers could also be used to detect damaging levels of particulates from the large, widespread ash clouds caused by erupting volcanos. In a different application, the use of laser bar code readers has been suggested as a means for automated optical identification of aircraft near to or on the ground. This could assist controllers in keeping track of air traffic.

There are numerous other ways in which lasers and electro-optics are becoming enabling technologies in aviation. Commercial and military aircraft have used laser gyros for a number of years. In addition, optical fiber communications and optical sensors will play an increasing role in the aircraft of the future and provide pilots with more information than they currently have, which will be fused into intelligent optical displays. The infrastructure of quantum electronics has provided building blocks for many of these advances.

Ground Transportation

In ground transportation, optical techniques for detecting speeders are coming into use. A laser-based system can unambiguously discriminate a particular car and give a rapid and precise reading of its speed. The system is compact, efficient, lightweight, and safe to use.

Several programs are under way to develop "smart highways" and automated rail systems. Optical sensors identify the position of vehicles, and a systems analysis of a real-time optical display, suggesting alternate routes. Already used in a rudimentary form on highways, such systems would be dramatically improved by less expensive and more efficient display technologies.

Future generations of automobiles may use lasers and optical science in several places. Fiber optics has been used to demonstrate that the lighting systems required in an automobile all can be powered by a single light source. Inexpensive and localized laser radar systems may be used to assist in collision avoidance. Optical fiber sensors are being developed to monitor combustion within the cylinders of an automobile engine. Such sensors make it possible to optimize combustion within an operating automobile, improving efficiency. Although currently used only for research on improved automobile engines, someday such monitors may be implemented in all vehicles.

Laser-based CD systems are currently entering the market as sources of onboard maps for land vehicles and boats (and aircraft). When combined with inexpensive global positioning receivers, they can provide the operator of almost any vehicle reliable, detailed information on location and local terrain, road networks, and so on (Figure 2.14). Such systems are being tested, for example, in rental cars in Florida.

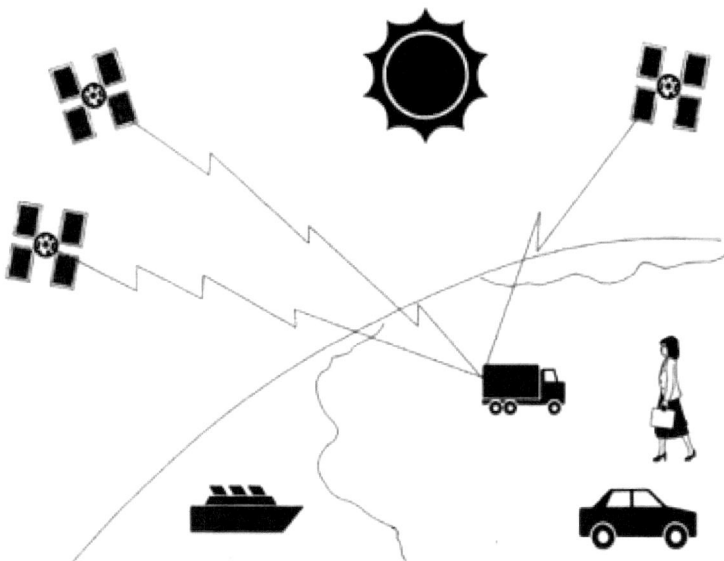

FIGURE 2.14 The extremely accurate timing made possible by atomic clocks has made the Global Positioning System a reality. An affordable receiver gets timing signals from satellites, and a built-in computer triangulates from the satellite locations to give locations on (or above) Earth within about one meter. Already these systems are coming into popular use for cars, trucks, boats, and planes.

AMO science is making a significant contribution to improvements in transportation systems and safety. The many new ideas currently under development suggest that this contribution will increase in the future

3

Education and Human Resources

SCIENCE EDUCATION

The nation's problems in science education have been described and documented in numerous reports and studies. The general level of scientific and technological literacy lags that of other developed nations and is inadequate for understanding and dealing rationally with scientific and technological issues and opportunities; precollege education and teacher preparation are in a state of disrepair; and despite universities and graduate programs that are the envy of the world, there is concern that the flow of talent into careers in science and engineering from all segments of society will be inadequate to meet future needs.

AMO science, through its focus on phenomena that are observed in the everyday world and its discoveries and technological developments, can contribute to science education at all levels. Optical wonders such as holography and lasers fascinate people of all ages and provide an excellent means to introduce them to the joy of science and to stimulate them to further scientific exploration and study. Once this interest established, AMO science can nurture it through the numerous opportunities it affords for practical hands-on experience and learning at all stages of development. Finally, training at the advanced level in AMO science produces flexible, capable scientists with knowledge and skills that apply to a wide range of disciplines and needs.

Developments in display and reprographic technology made possible by AMO science are having an important impact in the classroom. Devices such as small and inexpensive copying machines, laser printers, video disks, and CD-ROMs find widespread, and frequently innovative, applications in classroom teaching. If inexpensive projection TVs that provide large full-color images become available, these too would be a valuable teaching tool.

K-12 Education

At an early age, children question the magic of the blue sky and red sunset and the colors of the rainbow. If properly encouraged, this curiosity about nature and the world around us can be extended naturally into a lifelong interest in science. Children who take for granted devices such as the laser printer or TV remote control can be stimulated to question how they function and discover their operating principles. Many of the underlying optical principles can be simply explained and demonstrated with experiments that are inexpensive and safe, work reliably, and are suitable for the grade-school student. Volunteers from the Optical Society of America have developed an Optics Discovery Kit, now sold to teachers and schoolchildren, to provide an opportunity for young students to undertake simple experiments and, through guided discovery, obtain insights into optical science and windows into other fields of science and engineering.

AMO science is well represented in the high school classroom. The study of atomic structure, spectra, and reactions is basic to the understanding of modern science. The availability of inexpensive lasers now makes possible a diverse array of physics demonstrations whose visual impact can capture the attention and interest of students and greatly facilitate learning. These demonstrations can be as simple as ray tracing in geometrical optics or as complex as creating holograms or fiber-optics communication systems. Lasers clearly fascinate students, as evidenced by their frequent use in science fair projects.

AMO science contributes to increasing the scientific and technical literacy of people of all ages because its many applications in areas ranging from high-technology weaponry to concerns about the ozone hole are often discussed in the media. This coverage exposes people to many scientific issues and questions and can stimulate further reading on these topics.

Advances in science and its applications are being made so rapidly that it is difficult for teachers to keep up with new developments. As a result, it is hard for them to communicate the excitement in science from the new concepts that arise. An important investment opportunity therefore lies in upgrading the curriculum of prospective and continuing teachers to acquaint them with exciting and rapidly moving research areas as well as to provide solid training in the fundamentals.

Undergraduate and Graduate Education

AMO science plays an important role in the education of science and engineering students at the college and university level. It forms an essential component of the course work undertaken by science and engineering students. In the instructional laboratory, it provides opportunities to carry out interesting and instructive experiments in active areas of scientific research (such as in optical pumping, holography, high-resolution spectroscopy, nonlinear optics, and even laser cooling) that hone their practical skills without the need for vast resources. The relatively small scale of equipment, manpower, and financial resources necessary to carry out research in AMO science allows it to be undertaken in a broad range of educational institutions. Undergraduate involvement in these research programs offers many future professional scientists their first encounter with the excitement of the search for new knowledge.

At the graduate level the small scale of AMO science requires that students get involved in many or all aspects of an experiment from vacuum system design to data analysis and interpretation, and from laser development to computer control of equipment (Figure 3.1). This experience provides students with a diverse array of practical skills and promotes independent thinking. Furthermore,

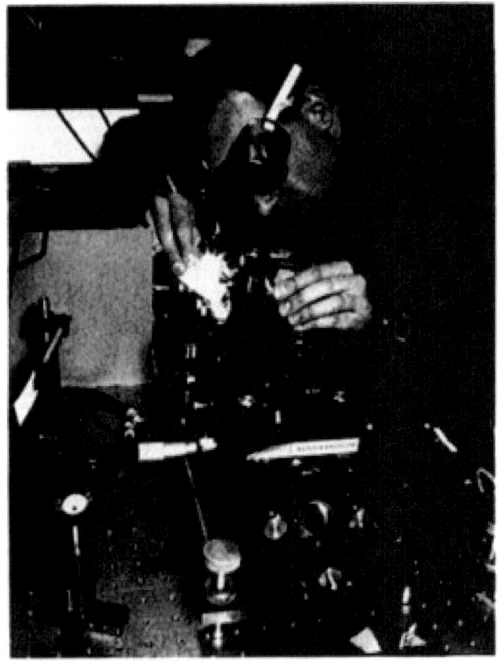

FIGURE 3.1 An AMO science graduate student at work on a small-scale experiment. (Courtesy of University of Virginia, Charlottesville.)

AMO science is unusual in that the same student may frequently both perform an experiment and carry out a theoretical analysis of the data. Of course, in many cases the theoretical analyses are sufficiently subtle and complex that more sophisticated theoretical skills are necessary, provoking collaboration between theorists and experimentalists.

Graduate training in AMO science, through its focus on phenomena that occur at energies characteristic of the world around us, provides students with skills and experience that have immediate application in addressing many of the problems facing the economic and technical vitality of the nation. AMO doctoral graduates play key roles in industry, government, and education and are among the professional men and women who are critically important to the future position of this nation in an increasingly competitive world. The broad training provided by AMO science also allows students to move rapidly into new areas and to attack complex, large-scale problems that require multidisciplinary input.

Graduates in AMO science are a valuable national resource. However, the future of science funding is in a state of flux. Thus it is sensible to reexamine, in the context of changing national needs and priorities, the training offered to AMO students. Degree programs should be reevaluated with a view to increased interdisciplinary emphasis; students and research programs will benefit from the differing perspectives brought to AMO science by researchers in chemistry, engineering, and physics departments. Students must be made more aware of the opportunities and needs in the many areas that are enabled by AMO science, possibly through cooperative programs with industry and government. Principal investigators and agencies should consider increasing the number of positions for postdoctoral fellows and decreasing the reliance on graduate students to carry out most of the research. In addition, the possible demand for students having master's degrees should be explored to see if the current emphasis on the doctoral degree in graduate education is warranted.

> AMO science is a valuable contributor to science education and literacy at all levels. For the nonspecialist, it provides an opportunity to stimulate further exploration of science and engineering and to demonstrate the direct and accessible correlation between experiment and theory. For the specialist, it provides hands-on training that imparts a diverse range of practical skills with immediate application in addressing many of the economic and technical problems of the nation.

HUMAN RESOURCES IN AMO SCIENCE

Ultimately, progress in any field is driven by the activities and capabilities of its human resource base. AMO science is a dynamic, interdisciplinary enterprise in which there is a significant movement of professionals into and out of the field. The data presented here are taken from the American Institute of Physics (AIP) *Employment Survey 1990* (AIP Report No. R-282.14, American Institute of Physics, New York, 1991), the Survey of Earned Doctorates (SED) containing data for the years 1961 to 1991,[1] and the survey conducted in 1992 by the Panel on the Future of Atomic, Molecular, and Optical Sciences (the FAMOS survey, discussed in Appendix D). These surveys all focus almost exclusively on PhD-level professionals and ignore the much larger number of individuals whose employment depends on applications of AMO science. Consequently, the data reflect the demographics of research employment. None of these surveys is exhaustive; they give different kinds of information; the data can only be regarded as approximate and are open to different interpretations. Taken together, they nevertheless provide some insight into human resources in AMO science.

Present Situation

For the year 1989 the SED estimates that there were 2,725 PhDs employed in AMO physics. The number employed in AMO chemistry is more difficult to estimate because the data contain no distinct category that corresponds to the present definition of AMO science. Nonetheless, the SED estimates that in 1989 4,666 PhDs were employed in physical chemistry, but this must represent an upper bound to the number employed in AMO chemistry as defined here. Taken together, the SED data suggest an upper limit of ~7,500 PhDs active in AMO science. The FAMOS survey indicated that ~59% of respondents received their degree in some area other than AMO physics. Thus, using the SED figure of 2,725 PhDs employed in AMO physics, the survey suggests that the number of PhD professionals actively engaged in AMO science is about 6,000 to 7,000 and does not appear to be changing rapidly. About half of these scientists work at universities and one-quarter in industry, and the remainder are employed primarily by the federal government or in government research laboratories.

The SED estimates that in 1989 there were 3,990 holders of PhDs in AMO physics and 12,476 holders of PhDs in physical chemistry employed in science. The data show that only 25% of the PhD professionals holding degrees in atomic and molecular physics were still active in that field, but that 91% of them remained employed in science or engineering. The retention in optical physics was considerably higher, with 50% of the degree holders remaining in that field. About 30% of the physical chemistry PhDs are employed in the field of their

---

[1] The Survey of Earned Doctorates is sponsored by five federal agencies—the National Science Foundation, National Institutes of Health, U.S. Department of Education, National Endowment for the Humanities, and U.S. Department of Agriculture—and is conducted by the National Research Council.

degree, but 23% work in fields other than chemistry, physics, or engineering. The large mobility out of AMO science demonstrates the broad applicability of AMO methods and techniques in a wide range of science and technology. The outward flow of professionals is balanced by a comparable influx from other disciplines: the FAMOS survey shows that only about 30% of the PhDs now working in AMO science actually have degrees in AMO physics. The distribution of PhD specialties of AMO professionals responding to the FAMOS survey is shown in Figure D.2 of Appendix D and confirms the interdisciplinary nature of AMO science and the career mobility within it.

The distribution of year of PhD attainment for respondents to the FAMOS survey is shown in Figure D.1 and is shaped by the retirement of individuals receiving their PhDs before 1955, by an increase in PhD employment between 1955 and 1970, and by a constant rate of new PhD employment between 1970 and the present, with some diffusion of the new degree holders into other fields after an initial employment in AMO science.

According to the SED, minority representation in AMO science is small but is characteristic of that in physics and chemistry as a whole. The group is about 93% male, 88% Caucasian, 10% Asian, 1% Hispanic, and 1% Black. About 80% are native U.S. citizens, and 12% are naturalized citizens. The only recent demographic trend has been that the fraction of AMO PhDs awarded to U.S. citizens has decreased from about 85% in 1961 to about 62% in 1991. Almost all of the foreign students come from Taiwan, the People's Republic of China, Korea, Hong Kong, Japan, India, and Canada. European, Middle Eastern, South American, Australian, and African students are present, but in insignificant numbers.

PhD Production and Initial Employment

The primary source of PhD AMO scientists in the United States is domestic graduate education. The SED shows that as a fraction of total physics degrees, the number of PhDs in AMO physics has remained essentially constant at about 10% for the last 20 years. U.S. institutions granted 160 AMO physics PhDs in 1991, and this annual output has been relatively constant, varying between a high of 182 in 1970 and a low of 115 in 1985. The time required to obtain a PhD has been constant at about 6 years since before 1970. The only noteworthy change has been a steady increase in the fraction of degrees in optical physics, growing from about 10% of the AMO total in 1969 to about 50% in 1991. During the same period, the fraction of physical chemistry degrees declined from 28% to 18% of all chemistry PhDs, with the absolute number of physical chemistry PhDs granted decreasing from 506 in 1969 to 408 in 1991.

The 1991 AIP *Graduate Student Survey 1989-1990* (AIP Report No. R-207.23, American Institute of Physics, New York) reveals that, after graduate school, about two-thirds of the atomic and molecular physics graduates took postdoctoral positions, while the remainder went into potentially permanent positions. This initial employment distribution is characteristic of all physics graduates as a group. Three-quarters of the optical physics graduates, however, secured long-term positions, indicating a relatively high demand for personnel in the optical area. Of physics graduate students seeking potentially permanent employment, those in AMO science had among the highest probability of receiving at least one job offer. The FAMOS survey data indicate that about 5 to 10% of the AMO PhD work force is in temporary positions and point to a relatively high incidence of postdoctoral employment: more than half of the research groups in universities and government laboratories contain one or more postdoctoral associates and have hired postdoctoral researchers within the last 3 years. Most of the FAMOS survey respondents in academe and government (58% and 67%, respectively) reported regular interaction with postdoctoral associates.

Most of the initial positions held by AMO PhDs in academe and government are postdoctoral, while potentially permanent initial positions are primarily in industry. (The FAMOS survey shows that industry employs relatively few postdoctoral researchers.) The SED reports that the distribution of new AMO PhDs' first employment is about 65% industrial, 20% university, and 15% government and that this distribution has been approximately constant for the last 10 years. The FAMOS survey, however, indicates that only about 37% of recent AMO PhDs are employed in industry. Taken together, these data may imply that at the time of their initial employment, new PhDs often take industrial positions in non-AMO areas; but a reduction in hiring of PhD scientists by industry would also be consistent with the data. Combining the AIP and FAMOS survey data suggests that about half of all AMO PhDs are initially employed outside AMO science. This conclusion is consistent with the AIP employment survey, which found that only 56% of PhDs in atomic and molecular physics remained in the field for their postdoctoral work. This healthy flow of AMO expertise, predominantly into the industrial community, promotes technology transfer and reemphasizes the value of the skills and knowledge imparted to the student via an education in AMO science.

Future Needs

It is reasonable to anticipate a constant level of postdoctoral positions in the near term. The FAMOS survey shows that those groups that now employ postdoctoral researchers also anticipate hiring them in the next 3 years. The situation is more negative for near-term permanent employment: the survey reports that about 50% of the respondents' departments and research groups had attempted to hire a permanent employee in the last 3 years, but that only about 40% of the groups and departments intend to make permanent hires in the next 3 years. The near-term employment picture is strongly influenced both by the economic downturn of the 1990s and by changes in the defense and aerospace industries and in the national laboratories that, traditionally, have been performing defense and energy research. Thus for the next few years, more AMO graduates will probably find employment in non-AMO fields, and the numbers in the "postdoctoral pool" may increase.

The long-term future human resources needs will derive from needs within the AMO community and from demands of industry, government, and academe outside of AMO science. The employment within AMO science has been constant since 1970. Even if most academic positions are refilled as their present holders retire, this demand is relatively small and might absorb 10 to 15% of the projected future PhD output. The demand for AMO graduates in other areas is more difficult to predict. Many industries that have typically hired AMO graduates, such as the microelectronics manufacturing, aerospace, and defense industries, are all now under stress and face an uncertain future. The outlook is somewhat brighter for AMO PhDs in other industries, such as communications, environmental monitoring and control, and medical instrumentation. As noted elsewhere in this report, future national economic growth may hinge on increased research and development activity in AMO science because of its many contributions to industry. If this increase occurs, there will be an augmented need for AMO graduates; if not, the demand will probably follow the recent historical trend of constancy.

> The historical and demographic data show that PhDs in AMO science undertake a wide range of occupations, with many working in areas enabled by AMO science, but few leaving science. AMO graduates are readily adaptable to occupational mobility, perhaps because the small scale of most AMO projects requires students to become intimately involved with all aspects of the project, including planning, design, execution, and data analysis and presentation. PhD graduates leaving the research laboratory for employment in industry are themselves a most effective vehicle for technology transfer. They carry with them internal knowledge of new science and techniques and are able to apply these effectively in an industrial environment. The industries that employ AMO scientists are those that have contributed significantly to recent economic growth in the United States and are most needed to sustain its economic health.

4

Funding and Infrastructure for Research and Development in AMO Science

AMO science is typically "small" science undertaken by individual investigators or small groups. In a few applications, such as laser fusion and laser isotope separation, AMO science becomes "big" science, requiring large facilities and large research teams. The national investment in AMO science has helped produce a vigorous and productive science that creates and supports important technologies. AMO science continues to advance with the introduction of new ideas and the invention of new techniques.

RESOURCES

This study comes at a time of considerable discussion and uncertainty about the future of funding for all physical sciences. Because of its vitality as a basic science and its relevance to a wide range of problems, AMO science currently is supported by a large number of agencies and is actively pursued by many academic, federal, and industrial laboratories. But the assumptions and policies of the past decades are undergoing reexamination, driven largely by the end of the Cold War and increasing concerns about national competitiveness in a global economy. Federal support for much of physical science has been based either explicitly or implicitly on the impact of earlier research on defense-related technologies during and after World War II. The reexamination of the rationale for funding for research and development (R&D) and the reality of limited federal resources present AMO science with challenging and potentially difficult times. A significant fraction, perhaps one-half, of AMO R&D funding has been defense-related. Defense-related R&D grew rapidly in all areas in the first half of the last decade to the point that it accounted for over two-thirds of all federally funded R&D, but then it leveled off, and it is now declining. The impact of this reduction on basic research is not clear because basic research accounts for only a small fraction of all defense R&D. Although the present debate concerning the reasons for support of R&D makes this a difficult time for science, it also provides an opportunity to place AMO science appropriately within the context of national needs and priorities. This is an especially good opportunity for AMO science because of its broad economic and societal impact.

Consideration of the relative roles of government and private industry in promoting and supporting long-range R&D is an important component of the national debate. Methods for improving technology transfer and industrial competitiveness, and the roles of individual federal agencies and laboratories in such activities, are being analyzed and discussed. Over the past 30 years, private industry has accounted for an increasing share of all R&D in the country (rising from ~33% in 1960 to ~51% in 1991); much of this increase is in health-related sciences. In AMO science, several industrial laboratories have played important and highly visible roles in basic as well as applied research, but AMO research in these laboratories is now declining. Difficult and uncertain economic times, changing corporate structures and markets, and changing philosophies regarding basic research have resulted in reductions in the amount of long-range R&D carried out in industrial research laboratories, even though much of this work was of high quality and visible. The reductions in R&D activities appear to be broad-based. The 1992 edition of the annual R&D trends survey conducted by the Industrial Research Institute exhibits many indicators of reductions in industrial R&D. The major finding of that survey is that "1993 will see the recession continuing for industrial R&D in the United States." Of the 141 companies responding to the survey, 36% expect decreases in R&D capital spending, while only 20% plan increases; 40% expect decreases in hiring of new graduates, and only 10% plan to increase hiring; and 32% plan decreases in "directed basic research," while only 12% plan increases. The federal government must of necessity carry the major responsibility for supporting basic science, but there are obvious advantages in having active basic science along with applied R&D programs in industrial laboratories to promote links between the science and the technologies that derive from it.

In an environment shaped by increasing concern about federal budget deficits, national priorities must be set and difficult decisions must be made. Setting priorities is a natural and necessary part of any budgeting process. In fields of science where large facilities, such as telescopes or accelerators, are central, formal priority setting is an absolutely necessary and accepted part of the process. In AMO science, where much of the science is done by single investigators or small groups, the pattern has been to let the individual investigators and the peer review process determine the directions of research. The AMO science community believes strongly in an emphasis on individual creativity and in the effectiveness of the individual investigator. Indeed, respondents to the FAMOS survey questionnaire indicated that the highest priority should be single-investigator, small-scale programs (see Appendix D). Competitively funded, peer-reviewed, small-scale science promotes a kind of intellectual capitalism; individual scientists choose what to work on and promote their ideas and results. Their success is ultimately determined in the marketplace of ideas and by their peers, who help make decisions about funding, publications, and other kinds of recognition. This system has produced remarkable results. It is, however, not too difficult to identify broadly defined areas of promise, and it is possible to identify (as is done in this report) which areas of AMO science are particularly relevant to specific national needs, such as energy or information technology and communications.

In connection with this study, an attempt has been made to carry out a systematic and comprehensive survey of the financial resources invested in AMO science. Two measures were used, one generated from the budgets of the agencies and laboratories supporting AMO science and the other from numbers supplied by respondents to the FAMOS survey questionnaire. The survey of agency budgets is complicated by the large number of agencies, and divisions within agencies, that support AMO science. In part, support is distributed within and across agencies because AMO is an interdisciplinary science that cuts across traditional boundaries of physics, chemistry, and engineering. Also, the very nature of the field makes it difficult to draw precise boundaries, but boundaries must be defined in order to generate consistent numbers. As described in the preface, the boundaries are least clear in the molecular and optical areas. In this report the panel defines AMO science according to the list in the Research Specialties Directory generated for use with the questionnaire and reproduced in Appendix D. This definition is much more inclusive than that used in the "Atomic, Molecular, and Optical Physics" section of the 1986 National Research Council (NRC) report *Physics Through the 1990s* (National Academy Press, Washington, D.C., 1986). This change in definitions, while necessary and appropriate, means that one cannot compare the funding levels reported here directly with those presented in the earlier study. Here the panel concentrates on generating a snapshot of AMO funding for fiscal year (FY) 1991 (the responses to the questionnaire give numbers for only that year), together with, where possible, federal funding levels for AMO science for FY89, FY90, and FY91. This time period is not long but should reveal any substantial recent reallocations of resources.

It is difficult to estimate the total size of the national effort in AMO science because this depends on how the boundaries are drawn, both the boundaries of the field and the boundary between research and development. For example, at Lawrence Livermore National Laboratory the support for basic AMO science is approximately $6M per year, but the entire laser-related R&D effort, including laser fusion and laser isotope separation, totals about $200M per year. One measure of the size of the field is the total amount of resources quoted by the respondents to the FAMOS questionnaire. As noted previously, recipients of the questionnaire were given the Research Specialties Directory (taken as the working definition of AMO science) and were asked to identify their areas of activity and to give annual support levels. The respondents listed total annual support grants, contracts, and donations for AMO R&D of ~$610M per year. It is possible there is some double counting, but respondents were asked to quote only their individual share of any multiinvestigator support. This figure does not include all R&D efforts in AMO science, for example, the large laser programs at Livermore or activities in certain industrial laboratories. On the basis of the survey data, it appears that the total annual R&D funding for AMO science, broadly defined, is probably on the order of $1B.

Federal Funding for Research in AMO Science

The principal sources of R&D support for AMO science are the National Science Foundation (NSF), Department of Energy (DOE), Department of Defense (DOD), National Aeronautics and Space Administration (NASA), National Institute of Standards and Technology (NIST), and private industry. Data from the FAMOS questionnaire suggest that ~75% of the total support is provided by federal agencies, with ~90% of this from the first four agencies listed above. Industrial contracts and corporate donations account for ~15% of the total support. The different federal agencies operate largely independently and have differing missions. In this study the most careful examination was done of agencies whose principal role is awarding grants and contracts because for these agencies the data are the most accessible. It is difficult to separate AMO science as defined here from other science and engineering in industrial organizations (where the required data are frequently not available) and in many federal laboratories, but the effort has been made in representative cases.

In the present study the field definition provided by the Research Specialties Directory was the most carefully applied in determining the support level from NSF and DOE. For these agencies, abstracts of funded proposals were made available for the past several years, and there were individually examined to determine if they lay within the field definition. In agencies other than NSF and DOE, figures for the field were, for the most part, generated by program managers within the agencies. The total annual federal funding of grants and contracts for basic research in AMO science is on the order of $100M.

National Science Foundation

The primary mission of NSF is to promote the progress of science and engineering, and it is the largest source of support for basic AMO science. NSF relies on the merit review system to determine which grants to fund. Support for AMO science is provided by six disciplinary divisions: Astronomy, Atmospheric

TABLE 4.1 Annual Federal Agency Funding Levels for Basic AMO Science in Then-Current-Year Dollars for FY89 Through FY91 (in thousands of dollars)

Source	Period	Experimental			Theoretical			Total
		Atomic	Molecular	Optical	Atomic	Molecular	Optical	
National Science Foundation	FY89	7,905	16,853	4,905	2,126	4,000	1,233	37,022
	FY90	8,404	16,710	5,884	2,465	6,254	909	40,626
	FY91	8,949	19,340	7,004	3,233	5,466	1,315	45,307
Department of Energy	FY89	8,405	8,628	2,558	2,227	2,733	316	24,867
	FY90	7,773	10,386	2,377	1,965	3,816	267	26,584
	FY91	8,698	11,426	2,339	2,252	3,005	296	28,016
Air Force Office of Scientific Research	FY89	1,161	6,106	12,560	54	1,708	387	21,976
	FY90	1,135	5,561	12,061	54	2,168	389	21,368
	FY91	1,118	5,260	10,786	53	2,027	369	19,613
Office of Naval Research	FY89	800	1,300	2,850	125	1,825	500	7,400
	FY90	680	1,530	2,550	210	1,885	460	7,315
	FY91	1,050	1,160	2,850	230	1,965	590	7,845
Army Research Office	FY89	700	710	2,850	60	0	0	4,320
	FY90	410	740	2,290	87	0	0	3,527
	FY91	660	735	2,180	39	0	0	3,614
TOTAL	FY89	18,971	33,597	25,723	4,592	10,266	2,436	95,585
	FY90	18,402	34,927	25,162	4,781	14,123	2,025	99,420
	FY91	20,475	37,921	25,159	5,807	12,463	2,570	104,395

Sciences, Chemistry, Engineering, Materials Research, and Physics. The largest sources of support are the Physics and Chemistry divisions, each with more than 40% of the total. Abstracts for the years FY87 to FY91 from program elements with possible AMO content in all six divisions were examined and classified according to whether the subject was atomic, molecular, or optical science and whether the work was experimental or theoretical. Not surprisingly, a substantial number of individual research programs involved work in more than one area, and the classification into atomic, molecular, and optical is thus somewhat arbitrary. Nevertheless, the classification gives some measure of the way support is distributed. The total support for FY89 to FY91 is shown in Table 4.1. Because the abstracts are recorded only in the first year of multiple-year grants, the present methodology may result in some underreporting of the earlier years, but the errors should be small. The increases in funding from FY89 to FY91 are larger than inflation. In constant FY89 dollars, the totals for the 3 years FY89 to FY91 are approximately $37M, $38M, and $41M, respectively, an 11% increase. While the level of support has been growing, there also is an increasing tendency for the support to be restricted to certain strategic needs, such as the Federal Coordinating Council for Science, Engineering, and Technology (FCCSET) initiatives.

It is useful to compare the figures in Table 4.1 with those obtained from responses to the FAMOS questionnaire. It seems likely that the rate of response to the questionnaire was highest among funded university and college faculty. Almost all (93%) of the NSF support goes to research programs at universities and colleges. If it is assumed, for the sake of argument, that the response rate is close to 100% for this category, comparing the total for FY91 above to the total reported by respondents to the questionnaire provides a test comparison of the two methods. The total FY91 NSF support reported by respondents to the questionnaire is $59M, compared to the $45.3M shown in Table 4.1. This comparison suggests that more people identify themselves as doing AMO science than are included in the present analysis of NSF funding.

The average size for all NSF grants (theory and experiment) included in Table 4.1 is $100K. This is compared with an average NSF grant of $86K reported by respondents to the questionnaire and with an average of $92K for FY83 given in *Physics Through the 1990s*. In constant FY83 dollars, the average FY91 grant of $100K is only $75K. This erosion in grant size is alarming but is recognized by the program managers. The Physics Division has increased the average annual award level for experimental AMO physics from $120K in FY91 to $156K in FY92.

Department of Energy

The Department of Energy is another major source of support for AMO science. DOE's mission is to develop efficient, dependable, and environmentally acceptable techniques for the transformation, delivery, and use of energy and to help ensure national security. The department supports AMO science because of its relevance to energy research and its applicability to defense programs. The DOE figures included in Table 4.1 were derived from a detailed review of abstracts from the divisions of Chemical Sciences, Materials Sciences, Advanced Energy Projects, and Engineering and Geosciences in the Office of Basic Energy Sciences (OBES) and from information supplied by the Office of Fusion Energy (OFE) and the Office of Health and Environmental Research (OHER). The increase in support evident in Table 4.1 from FY89 to FY91 is slightly greater than inflation. In constant FY89 dollars the FY91 total represents an increase of approximately 3%.

The figures in Table 4.1 include basic AMO science efforts at Ames, Argonne, Brookhaven, Lawrence Berkeley, and Oak Ridge national laboratories that are supported from the same budgets as those that support research at universities and allied institutes. The national laboratories account for approximately 50% of the support. Other DOE laboratories, such as Los Alamos and Lawrence Livermore, receive block funding from DOE and are not included in Table 4.1. For comparison, the annual totals for basic AMO science at the Los Alamos and Lawrence Livermore national laboratories are estimated to be $5,250K and $6,350K, respectively, while large applied AMO programs such as inertial confinement fusion have budgets totaling a few hundred million dollars. The table does not include grants and contracts made to national laboratories and universities by OHER. Total annual OHER support through FY91 for research in AMO science was in excess of $3M.

As with NSF, respondents to the FAMOS questionnaire at colleges and universities report significantly more support, $26M, from DOE than suggested by the ~$13M of university FY91 support that is part of the ~$28M indicated in Table 4.1. This comparison may be considerably affected by those university researchers who have joint appointments or carry out their research at the DOE laboratories.

Department of Defense Research Offices

The Department of Defense supports AMO science through the Air Force Office of Scientific Research (AFOSR), the Army Research Office (ARO), and the Office of Naval Research (ONR). In addition, AMO science or AMO-related work is carried out at many DOD laboratories.

The funding of AMO science through AFOSR comes largely from the Physics and Electronics Directorate and from the Chemistry and Materials Science Directorate. Proposals to AFOSR are reviewed by an Air Force laboratory scientist and must overlap the mission of an Air Force laboratory. AFOSR support for AMO science is summarized in Table 4.1. The figures in Table 4.1 include estimated University Research Initiative support of about $700K per year in molecular experimental work and $1,000K per year in optical experimental work. The table does not include support of approximately $7M per year for basic research (6.1 funds) in AMO science carried out in the Air Force laboratories. The downward trend in research support evident in Table 4.1 is more dramatic when inflation is taken into account.

As the first agency empowered to make federal research contracts in the post-World War II reconstruction period, ONR, established by the Vinson Act of 1946, has had the longest continuing commitment to AMO science. Its program of support remains a vital component in the overall health of the field. In recent years, about half of the support of basic AMO science has been in base programs of ongoing research, and half has been provided through accelerated research initiatives that support defined target areas for 5-year periods. The base programs also involve support of chosen thrust areas but may provide support for periods longer than 5 years. The largest component of support for AMO science comes from three divisions in ONR: Physics, Chemistry, and Mechanics. The numbers in Table 4.1 do not include support from the divisions of Materials, Electronics, and Biology, which may amount to something on the order of $2M per year. The table also does not include the approximately $10M per year of support from ONR for basic AMO science at the Naval Research Laboratory.

The Army Research Office supports AMO science as part of its mission to provide the requisite science base for the Army's technological needs. ARO works to articulate the technical problems and performance goals of the Army to the scientific community to promote and acquire science that is specific to Army requirements. The ARO research program seeks to seed scientific and technological work that promises major advances over current technologies. ARO support is summarized in Table 4.1 and includes a large University Research Initiative program that is scheduled to be reduced by $800K per year in FY93.

National Aeronautics and Space Administration

To support its multiple missions in planetary and space sciences and exploration, NASA requires a wide range of information and results from AMO science, some of which is funded directly by NASA. In FY92, four divisions (Solar System Exploration, Astrophysics, Space Physics, and Earth Science) provided a total of approximately $7.3M of support for AMO science. This total consisted of $6,173K for experimental work (atomic, $1,967K; molecular, $3,468K; optical, $738K) and $1,128K for theoretical work (atomic, $1,057K; molecular, $71K).

Total Funding from Federal Grants and Contracts

The estimated total support for basic research in AMO science from federal grants and contracts is shown in Table 4.1. These figures include individual grant and contract support from the DOE and AFOSR for basic AMO science at certain federal laboratories. For FY91 the total amount of that support was approximately $20M. Thus, support of AMO science in nonfederal laboratories amounted to ~$84M in FY91.

TABLE 4.2 Total Annual Federal Agency Funding Levels for Basic AMO Science in Constant 1989 Dollars for FY89 Through FY91 (in thousands of dollars)

Period	Experimental			Theoretical			Total
	Atomic	Molecular	Optical	Atomic	Molecular	Optical	
FY89	18,971	33,597	25,723	4,592	10,266	2,436	95,585
FY90	17,459	33,138	23,873	4,536	13,399	1,921	94,326
FY91	18,641	34,524	22,905	5,287	11,347	2,340	95,044

While the totals in then-current dollars in Table 4.1 show increases each year, these increases have not quite kept up with inflation. This finding is consistent with results from the FAMOS questionnaire, where respondents indicated that funding has decreased somewhat during the last 5 years when inflation is taken into account. Table 4.2 shows total federal agency funding in constant FY89 dollars.

The average AMO grant size reported by university-based respondents to the questionnaire is $81K. Even in constant dollars, this is less than the $96K (in FY83) average grant size decried in *Physics Through the 1990s* as too small to sustain an active program. In constant FY83 dollars the present $81K average is only $61K, corresponding to a 36% reduction in the size of a typical grant. Grant sizes reported by respondents from industry and federally funded laboratories and government are significantly larger. More than half the respondents to the AMO questionnaire who receive direct support obtain that support from two or more grants. Despite the funding difficulties, however, the field has remained vital, as evidenced by the number of new and exciting discoveries in the last 10 years.

Given the modest average grant size, it is interesting to note that the respondents to the questionnaire, by a 2 to 1 margin, reject the idea of increasing the funding for the strongest programs at the expense of the total number of programs supported. But given modest increases in support, respondents in all employment sectors identified support for single-investigator, small-scale programs as their number-one priority. Funding for young investigators and equipment, two other areas often identified as needing increased emphasis, were also given high priority. On the other hand, in comparisons with other countries, respondents rated the United States as strong in equipment (and very strong in innovation).

Federal Laboratories

There are more than 700 federal laboratories supported by 14 federal agencies, and 190 of these laboratories employ 50 or more scientific and technical personnel. The laboratories involved in AMO science are those funded by DOE, DOD, NASA, and the Department of Commerce (DOC), which supports NIST.

National Institute of Standards and Technology

AMO science is of major importance to the missions of the National Institute of Standards and Technology. The total FY92 funding for AMO-related science in the NIST Physics Laboratory was $32M. Seventy percent of this came from NIST's own budget, and 30% was in the form of grants and contracts from other governmental agencies and reimbursable services. The effort was divided into four thrusts: Atomic and Molecular Structure and Dynamics, Physics of Surfaces and Materials, Physics of Electronics and Magnetics, and Optical and Laser Physics and Technology. Of the $32M, $11.2M was for basic research, of which three-quarters was for atomic and molecular science and one-quarter for optical science.

Department of Energy Laboratories

The changes in the economic and political climates in the world are anticipated to precipitate changes in the functions of the various DOE laboratories, especially the weapons laboratories. Studies are in progress, and laboratories are examining differing scenarios for their future. It is too early to predict the outcome, but it appears change will come, both a change in missions and perhaps some downsizing of laboratories. The laboratories are promoting programs to address economic competitiveness, energy, environment, health, and science education.

The DOE laboratories with significant components of AMO-related research include Ames, Argonne, Brookhaven, Lawrence Berkeley, Lawrence Livermore, Los Alamos, Oak Ridge, and Sandia national laboratories. The total support for basic AMO science at these laboratories is on the order of $30M. However, as already noted, the total AMO-related budgets, which include large applied AMO programs such as inertial confinement fusion, reach a few hundred million dollars. The laboratories also support large multidisciplinary facilities, such as the synchrotron light sources at Berkeley, Brookhaven, and Stanford, the Los Alamos Meson Physics Facility (now slated for shutdown), and the accelerators at Argonne, Lawrence Berkeley, and Oak Ridge, which are used in part for AMO science.

Department of Defense Laboratories

AMO science is also carried out at a number of DOD laboratories, including both Air Force and Navy laboratories. The Air Force laboratories with significant AMO science include the Geophysics and Astronautics divisions of the Phillips Laboratory and the Aeropropulsion Division of the Wright Laboratory. Table 4.1 does not include the approximately $5M per year in molecular science and approximately $2M per year in optical science provided by AFOSR for basic research at Air Force laboratories. A large amount of basic and applied AMO science is carried out at the Naval Research Laboratory, which is supported by the Office of Naval Research. The support for basic research in AMO science at NRL amounts to about $10M per year ($0.5M for atomic, $3.5M for molecular, and $6M for optical science). Applied AMO science is pursued at other non-ONR Navy laboratories.

INFRASTRUCTURE AND FACILITIES

The success of any field of science depends on an infrastructure within which new knowledge can be created. The achievements of AMO science noted elsewhere in this report reveal that AMO science indeed has had an effective supporting infrastructure and that the field is generally robust and healthy. The purpose of this section is to briefly examine that infrastructure and to discuss the future directions of its evolution.

The Single Investigator

There is no substitute for individual creativity and the ability to follow creative ideas to realization. This statement captures much of the spirit of "single-investigator science," which dominates the manner in which AMO science is conducted and is believed by most people in the field to be the most effective system. Most of the recent conceptual breakthroughs discussed in this report have resulted from this approach. This approach is tempered, however, by the realization that synergistic interchange with colleagues of like interest is beneficial and that many experiments can be undertaken only by using large, costly facilities. Furthermore, given the diversity of AMO science, development beyond an original idea often needs cross-disciplinary input and targeted research. Thus, AMO science is carried out in a balanced environment, where single-investigator, small science is the most prevalent, but centers and institutes, special facilities, mission-oriented research groups, and expensive instrument clusters fill important roles in bringing about the success that AMO enjoys.

The term "single investigator in the university setting" means a single senior investigator who leads and trains a group of graduate and undergraduate students and postdoctoral research associates. The senior investigator is typically the principal investigator on one or more grants or contracts with the supporting agencies in the federal government. In the national laboratory and industrial setting, the term "single investigator" usually refers to one to three senior scientists working together on a project, where again one of them may be the principal investigator on the funds obtained internally or through external contracts.

Letters received from about 100 leaders in the AMO science community, in response to a solicitation of their comments, overwhelmingly emphasized the importance of the single investigator, while expressing the need for the balanced approach present now in the field. As shown in Figure D.21 in Appendix D, respondents to the FAMOS questionnaire from industry, academe, and the national laboratories all believe that if only a limited amount of additional funding is available for the field, single-investigator, small-scale programs should receive the highest priority. However, it is clear from the responses that a balanced approach is important. Almost 75% of questionnaire respondents (see Appendix D, Figure D.8) are involved in collaborative research, and of those so involved, about 60% said their work is cross-disciplinary.

The single-investigator approach is not without its perceived problems, however. Among those pointed out by letter respondents was the observation that some single-investigator AMO academic groups may be too large, raising concern about whether the students could receive optimal education and training. Of course, whether the resources of the field allocated to such groups was optimally or fairly spent also received comment. Suggestions were made that group size limitations (e.g., to four or five students) might be imposed by the agencies. About 20% of university groups reporting in this survey indicated group sizes of more than five students.

The small-scale nature of AMO projects may place additional burdens on the AMO community for carrying out proper merit (peer) review of proposals. Nevertheless, opinions expressed in the survey and letters and by members of the panel all support strong, effective, and universally applied merit review for all proposed projects.

Centers and Institutes

Complementing the single-investigator approach to research in AMO science, a number of centers and institutes have been established. Most numerous of these are collections of individuals and departmental subgroups at various universities that have formed "centers" to promote communication, synergistic interactions across and within disciplinary and departmental lines, and visibility both within and without the university and to boost economies in infrastructure. Generally, these centers do not receive central funding from any source, save perhaps some infrastructure funds from their home institutions, but rather operate as a collection of single-investigator groups. Because new centers are being formed frequently, it is not sensible to list these entities individually.

Two major institutes have been established especially for optics, the Institute of Optics at the University of Rochester and the Optical Sciences Center at the University of Arizona. Both were established to meet a recognized national need for optical research and training in optical sciences. Both are degree-granting departments at their respective universities, and together they account for a major portion of the degrees in optics granted in this country.

Centrally funded cross-disciplinary institutes have taken a prominent position in the field as well. One of the first of these (founded in 1962), which has served as a prototype for others, is the Joint Institute for Laboratory Astrophysics (JILA) in Boulder, Colorado. Important to the AMO community as a whole is the institute's competitive Visiting Fellows Program, which supports about 10 senior visitors each year for a period of 6 months to 1 year, stimulating extensive cross fertilization within the field. With experimental and theoretical programs in atomic physics, molecular physics, and optical physics, and also in precision measurement, the institute represents a fusion of all the disciplines of AMO science discussed in this report, including a number of the applied areas. Another centrally funded institute that bridges many of the boundaries between different specialties is the recently formed Institute for Theoretical Atomic and Molecular Physics (ITAMP) at Harvard University. Responding to the AMO community's call for strengthening theory and theoretical training in the field, the institute was formed as a Harvard-Smithsonian partnership in 1989. The institute provides an intellectual center and meeting place where scientific interactions take place. The institute runs two or three workshops per year on topics of current interest and has programs for short- and long-term visits. In conjunction with the establishment of ITAMP, Harvard University created a new faculty position in its Department of Physics for a researcher in theoretical AMO science; that position has recently been filled at the rank of full professor.

Recently, in an effort to produce stronger coupling of the sciences to national goals and needs, the federal government, NSF and DOD in particular, has established a number of centers featuring cross-disciplinary approaches to the attainment of target goals. While recognizing the value of centers, letter responses from leaders in the field and respondents to the questionnaire (see Figure D.21) do not place a high priority on the establishment of new centers.

User Facilities

While AMO science is typified by small-scale, single-investigator research programs, responses to the questionnaire suggest that about 20% of experimental work and 29% of theoretical and computational work in the field are carried out at "user facilities" (see Figure D.9). However, respondents did not accord new user facilities high priority, though support for upkeep and use of existing facilities rated relatively high within the community (see Figure D.21). User facilities employed for AMO research include synchrotrons, accelerators, and storage rings. A discussion and listing of many of the facilities can be found in a recent report, *Future Research Opportunities in Atomic, Molecular, and Optical Physics* (PUB-5305, Lawrence Berkeley Laboratory (LBL), Berkeley, California, 1991). User facilities also include special high-power lasers, plasma machines, and high-speed computers.

The research value of synchrotron radiation was in large part explored and demonstrated in atomic physics studies in the 1960s, thus leading the way for other research communities to adopt synchrotrons in their work. The synchrotrons built recently or under construction in the United States have been primarily motivated by work in the condensed-matter and biological sciences. Nevertheless, synchrotrons continue to be a valuable tool for the study of atoms and molecules, because they provide a source of radiation continuously tunable from the infrared to the hard X-ray regime and their brightness allows the study of targets in the gas phase. In addition to the facilities listed in the LBL report, new facilities, including the Advanced Light Source at Berkeley and the Advanced Photon Source at the Argonne National Laboratory, are coming on-line soon, and both will support some AMO science.

A major issue concerning synchrotrons is that while construction of the basic facility is funded at a high level, funds for the building of beam lines are generally not included in the initial facility budgets, and these facilities may thus be underutilized. To use the machines, it is often necessary to raise substantial funds (costs may range from a few hundred thousand dollars to a few million dollars) to develop a beam line. This situation is a source of intense concern to AMO scientists who wish to use synchrotrons in their work, especially because funds for such relatively expensive new facilities are afforded a relatively low priority within the AMO community as a whole. However, two beam lines for AMO physics are funded and are currently under construction at the Advanced Light Source: an undulator beam for AMO gas-phase experiments with low-energy photons and a bend-magnet beam line for soft-X-ray experiments in AMO physics.

Accelerator facilities at Oak Ridge National Laboratory and at Kansas State University are dedicated to atomic collision studies at high energies. Other accelerators, such as the Los Alamos Meson Physics Facility (now under threat of closure), are maintained for other purposes but are also used for atomic physics. Again, the reader is referred to the 1991 LBL report for discussion of the facilities and the physics at these facilities.

Ion storage rings have become an important tool for studies of both highly charged atomic ions and molecular ions. These high-energy machines have recently been used for unique and sophisticated low-energy AMO experiments. Neither of two proposals for building storage rings in the United States that could accommodate AMO experiments managed to gain the necessary funding support. Thus, at this time there is no storage ring facility for AMO physics in the United States, and it is necessary for U.S. scientists to collaborate in experiments at European and Japanese facilities, where there are now five operative machines accommodating AMO experiments.

Research on the combustion process is one of the important areas outlined in this report. To facilitate this work, DOE maintains a user facility, the Combustion Research Facility at Sandia National Laboratories in Livermore, California ($4.3M annually). Current activities at the laboratory emphasize the development and application of new diagnostic techniques to the study of basic flame processes, research in fundamental chemistry in combustion, and analytical studies of reacting turbulent flow.

The most ambitious new facility currently being proposed is the Chemical Dynamics Research Laboratory at Berkeley, which would be built in connection with the Advanced Light Source. The proposed project currently is envisioned as an undulator beam line at the Advanced Light Source and conventional lasers, and, with necessary building and laboratory facilities, the cost is estimated to be approximately $66M. The laboratory would be dedicated to research programs in chemical dynamics, chemical kinetics, and spectroscopy of processes that are fundamental to the combustion of fossil fuels, goals that complement programs of the Sandia facility.

National Laboratories

Several studies are in progress to provide guidance on the future of the national laboratories, and it would not be productive for this panel to enter into this process. Nevertheless, it should be emphasized that there are AMO scientists in national laboratories who rank among the best in the world. Outstanding AMO theoretical and experimental groups have been established in these laboratories to address various national goals arising from defense and energy requirements. As these goals are shifted from defense-oriented to other priorities, care should be taken to ensure that these resources, the AMO scientists and infrastructure, are not lost to the nation.

Other Infrastructure Issues

Gathered here are a number of issues concerned with the infrastructure that have been called to the attention of the panel in town meetings, letters, and telephone calls and through other contacts with the panel.

Evolution of Subfields

Because AMO is an enabling science, a need may continue for data from a given subfield long past the time that the subfield is in the intellectual vanguard. Those doing the work may typically have been trained sometime in the past, but it is difficult to attract, support, and train new researchers in a subfield that might be considered "important but pedestrian." Thus, the health of valuable subfields may be neglected.

For example, the astrophysics community is strongly enabled by AMO science through data on atomic energy levels, wavelengths, lifetimes, and collision strengths. NASA supports some work in this area to obtain data that it directly needs, but there is little attention paid to the fact that essentially no new people are being trained in these areas. Quoting one letter respondent, "The training of students in this subfield has come to an end. The future is indeed bleak. We have discussed the possibility of hiring a young scientist to help carry on work in this area in our laboratory but have been unable to think of a single person with suitable training." The 1991 NRC report on astronomy, *The Decade of Discovery in Astronomy and Astrophysics* (National Academy Press, Washington, D.C.), also articulates this issue, especially in the second volume, *Working Papers*. Hence, the health of an enabling field can decline by benign neglect.

Another area where benign neglect has resulted in a community with aging experts and few newly trained practitioners is that of gaseous electronics. No agency accepts responsibility for this subfield, yet today it has renewed importance in the areas of plasma etching and manufacture.

An interagency committee charged to help coordinate research within and among the several agencies that depend on and are responsible for AMO science may alleviate recurrence of similar problems and may be able to redress some of the problems that have already occurred.

Theory

Independent of employment sector, field, or whether experimentalist or theoretician, nearly 70% of respondents to the questionnaire felt that the balance between theoretical and experimental work was about right, and among the remaining 30% there was a slight bias indicating too little theory. A 1987 NRC assessment of theoretical atomic physics in the United States, *The State of Theoretical Atomic, Molecular, and Optical Sciences in the U.S.* (National Academy Press, Washington, D.C.), found an unhealthy situation in terms of training of new people in the field and lack of practitioners in the top universities in the country. NSF responded with additional support, especially with respect to the establishment of ITAMP at Harvard, as mentioned above, and of a "minicenter" at JILA for theory. DOE also increased support somewhat. In the meantime, AMO theorists have participated in two summer institutes at the NSF Institute for Theoretical Physics in Santa Barbara. The theoretical AMO community formed an independent organization, Theoretical Atomic, Molecular and Optical Community (TAMOC), to promote communication. The response to the survey, the response of agencies that support AMO science, and broader participation in the theoretical community at large are all positive signs that the health of the discipline has improved. The panel would be remiss, however, not to point out that a number of leaders in the field continue to express, in their letters, concern about the long-term health of theoretical AMO science.

Instrumentation

Instrumentation continues to be a concern. Responses to the questionnaire (see Figure D.21) indicate that given a modest increase in funding, capital equipment would be among the highest priorities for its use. Letter respondents decried the state of equipment, citing much of it as significantly outdated—this for a field that leads in development of new measurement techniques and methods.

In addition to the need to redress the overall situation vis-à-vis instrumentation, two serious issues that need to be recognized and addressed deal with decreasing resources in the universities. The main opportunity a young scientist has to become established with equipment and apparatus for his or her research is through "start-up" monies received at the time of hiring. In universities, start-up funds range from the order of $100K to $1M, depending on the level of the person and the project(s). Agencies for some years have not generally made major start-up funds available, and the sciences have benefited enormously from the additional funds provided by university start-up packages. However, universities are under financial stress and are now less able and willing to allocate funds for this purpose. Where will the funding come from? There is a strong possibility that young people will have to rely more and more on existing setups and, thus, on old lines of research, if they are to have anything at all. A second-ary effect of the "start-up" problem is that it is extremely difficult for people to change research directions in mid-career even if they have excellent ideas.

Several of the agencies (NSF, DOD, and DOE) have sponsored programs specifically for upgrading equipment in universities. This support has helped enormously. However, the agency programs understandably have tried to increase the pool of money available for the upgrade by requiring matching funds from the operating institutions. Again, the declining financial situation in most universities has made matching funds more difficult, if not impossible, to obtain. AMO science is spread across a broad spectrum of universities, but researchers at many of these institutions express the opinion that those in the "wealthier" institutions benefit most from these matching programs. The upgrade of instrumentation remains a serious need in AMO science.

Academic Culture

The diversity of AMO science and its relevance and importance to many national goals make it naturally suited for interdisciplinary collaborations. Many university personnel complain that such collaborations are frowned upon in their academic departments, and activity in this direction often enters the ledger on the negative side in tenure decisions. It is to be hoped that such a narrow view of science and its role in the world will change in the future.

Postdoctoral Associates/Researchers and PhD Employment

In Chapter 3, it was noted that universities are producing more PhDs than the labor market is now employing. Inevitably, this means that the movement of AMO scientists into other areas will increase (which the panel views as a positive benefit), and the ratio of postdoctoral to graduate students in university research groups may increase. In the short term the mismatch between supply and demand creates great stress within the community and a serious morale problem. The nature of postdoctoral employment is a concern: it ranges from status as an advanced graduate student to that of an individual who conceives and executes research projects and directs graduate students in much the same capacity as a faculty member. As the pool of postdoctoral researchers grows, it is sensible to ask whether the infrastructure in AMO and other sciences creates and uses too many PhDs. The PhD salary level, in comparison with that in other professions requiring comparable educational preparation, suggests an oversupply. The composition and nature of the AMO science work force (that is, students, postdoctoral researchers/associates, professional scientists, technicians, and so on) is an important infrastructure issue affecting both the production of PhDs and their subsequent function in society.

Communication and Organization

The communication that is a vital part of any science endeavor is facilitated in AMO science by a number of professional organizations, in particular the American Institute of Physics, American Physical Society, American Chemical Society, Optical Society of America, Institute of Electrical and Electronic Engineers, and Society of Photographic and Instrumentation Engineers. The service of these societies and their AMO-related divisions should be emphasized. They provide the journals that are the primary vehicles for communicating research results. They sponsor meetings where face-to-face communications often spawn new ideas. They provide a forum and a means for recognition of accomplishment and thus help provide some of the motivation that drives scientists. The journals, the meetings, and the personal contacts are often taken for granted, and seldom are their economic value and importance to the infrastructure accounted for. The executive bodies of the divisions help provide a community voice to articulate achievements, needs, and problems to larger bodies. The panel believes that these organizations are an essential part of the AMO science culture in the United States.

The National Research Council's Committee on Atomic, Molecular, and Optical Sciences (CAMOS) provides a vital communications link between the community and the government agencies that have a vested interest in AMO science and provides a forum for discussing issues of mutual concern. It also provides an important communications link within and a voice for this diverse field, bringing together persons from the different areas, which tend to drift apart into their own divisions and societies. It provides a forum where problems and needs transcending individual subfields can be addressed. Through this committee, the AMO science community also has important ties to the Plasma Science Committee and the Solid State Sciences Committee of the NRC's Board on Physics and Astronomy, to the NRC's Board on Chemical Science and Technology, and to the NRC's National Materials Advisory Board.

5

Economic Impact of AMO Science

The panel could determine no accepted protocol for addressing the issue of the economic impact of AMO science. Economists appear not to agree among themselves on how to assess such impact or the "return on investment of research dollars." Therefore, within the human and financial resources available for this assessment, the panel has adopted a simple approach for viewing the impact, recognizing that more sophisticated perspectives might differ. Nonetheless, the approach taken demonstrates that AMO science has an impact on a sizable portion of the economy, and language used herein reflects this fact. For each of the areas considered, however, a number of different science and engineering disciplines also may have contributed substantially, and it is impossible in this limited assessment to define the role of each. There is thus no implication that AMO science *exclusively* enables all the commercial products or services listed, only that it is an important enabling factor.

Few if any fields of basic research directly produce marketable products or services. AMO science, however, plays a key role in many commercial areas; in some the role is obvious, in others it is not. In assessing the economic impact of AMO science, the panel can identify industries wherein AMO science plays an important enabling role, *that is, those in which the commercial activity would not exist without the contributions of AMO science* (e.g., fiber-optic communications and integrated circuit manufacturing). In a second class of commercial activities, the techniques and contributions of AMO science considerably enhance and add value to the product but are not essential to its existence (e.g., production of industrial chemicals, pharmaceuticals, and photographic equipment). A third class of commercial activities (primarily in the service sector) profits through

TABLE 5.1 Industries in Which AMO Science Is an Important Enabling Factor

Industry	Total 1991 Sales ($B)
Gas-discharge lighting	2.6
Semiconductor devices	27.8
Industrial and analytical equipment	
Laboratory instruments	13.4
Measuring and controlling instruments	10.5
Electronic information recording and storage (image data)	2.2
Information services on networks and electronic data storage media	10.2
Data processing and network services	35.6
Computer equipment and peripherals	
Supercomputers	1.5 (worldwide)
Mainframe	27.0 (worldwide)
Mid-range	26.0
Workstations	12.0
Personal computers	32.0
Portables	7.0 (worldwide)
Local area networks	8.0 (worldwide)
Magnetic storage equipment	3.0
Optical storage equipment	0.9
Laser printer equipment	6.3
Other printers	5.6
Telecommunications services	161.0
Radio communications and detection equipment	
Radar, sonar, LIDAR, navigation, infrared/ultraviolet, electronic warfare	31.6
Communications systems and equipment (fiber, microwave, satellite, land, marine, mobile)	17.5
Satellite communications equipment (ground-based)	1.3
Satellite communications equipment (satellites)	1.3
Commercial broadcasting equipment	1.8
Electronic kits, lasers, ultrasonic equipment particle accelerators, etc.	3.8
Fiber optics	1.7
Prerecorded music	
Compact disks	3.9
All other media	4.3
Consumer electronics (radio, TV, audio, etc.)	7.7

savings and productivity enhancements resulting from AMO science in a peripheral, but economically significant, way (e.g., savings in retail trade via reduction in labor cost and enhanced inventory control through bar code scanning at point-of-sale terminals). While the divisions among these three classes are not unambiguously drawn, Tables 5.1, 5.2, and 5.3 (data from *U.S. Industrial Outlook 1992: Business Forecasts for 350 Industries*, International Trade Administration, U.S. Department of Commerce, U.S. Government Printing Office, Washington, D.C., 1992) specify the economic activity in these three classes of U.S. industry in 1991.

Industry	Total 1991 Sales ($B)
Medical equipment	
X-ray equipment	2.1
Other electromedical	5.5
Health care directly derived from AMO	25.0 (estimated)
Nuclear magnetic resonance imaging—machine use	
Nuclear magnetic resonance imaging—radiology services	
X-ray tomography—machine use	
X-ray tomography—radiology services	
Microwave-based thermotherapy	
Laser-based diagnostic, surgical, and therapeutic procedures	
All procedures that involve internal inspection with fiber-optic devices including endoscopes, laparoscopes, etc.	
Laboratory analyses	
TOTAL	500.1

The economic impact of AMO science is significant. The GNP in 1991 was about $5,760B. Table 5.1 shows economic activity in which AMO science is an important enabling factor to be $500B. Table 5.2 totals the output of industries that are significantly enhanced by AMO science at about $634B, and Table 5.3 indicates an additional $114B of impact in other industries. Taken together, the tables indicate that AMO science significantly affects about 22% of the GNP. The areas of the GNP affected by AMO science are typically those having sizable annual growth rates, and one can therefore expect its importance to increase in future years.

AMO science is critical, for example, in the manufacture of integrated circuits and the products that such devices enable. Integrated circuit manufacturing relies heavily on advanced optics, surface analysis, and materials processing by lasers, ion beams, and plasmas. The interaction with AMO science will increase

TABLE 5.2 Industries in Which AMO Science Is a Secondary Enabling Factor That Provides Enhanced Product Value

Industry	Total 1991 Sales ($B)
Chemicals	
Organic chemicals	66.1
Inorganic chemicals	21.7
Paints and allied products	13.2
Adhesives and sealants	5.7
Fertilizers and pesticides	16.3
Plastic materials	32.2
Plastic and rubber products	70.0
Drugs	59.0
Refined petroleum products	136.3
Electronic components other than semiconductors	34.5
Industrial controls	6.9
Printing machinery	3.1
Private biotechnology R&D (77% in health care)	2.3
Government-sponsored health care research	4.0
Advanced materials	
Biotechnological materials	6.0 (estimated)
Ceramics	10.0 (estimated)
Powder metallurgical materials	2.0
Robotic equipment	0.5
Aerospace	
Civilian aircraft	27.9
Military aircraft	16.0
Aircraft engines	23.7
Aircraft parts and equipment	22.2
Guided missiles and space vehicles	21.7
Other space-related equipment	4.9
Photographic equipment and supplies	23.2
Microfilm image recording and storage	1.4
Printing capital equipment	3.0
TOTAL	633.8

in the future as devices become small enough that quantum effects become important.

TABLE 5.3 Large Industries in Which AMO-Related Devices Have a Peripheral Impact with a Significant Dollar Value

Industry	Total 1991 Sales ($B)	Estimated Impact of AMO Science ($B)
Printing and publishing	161.0	
Cost savings through laser platemaking, optical scanning, computer composition, laser printing for lithographic masters		16.0
Automobiles	133.0	
Electronics		13.0
Emission monitoring and control		1.0
Contributions to manufacturing process		1.0
Retail sales	1,900.0	
Savings in inventory control and management through bar code scanning		5.0
Savings in labor through bar code scanning		20.0
Health care	738.0	
Savings due to computer and electronic information management		5.0
Trucking	257.0	
Computer scheduling and satellite tracking of vehicles		3.0
Insurance, accounting, leasing, management and legal services	910.0	
Savings due to computer and electronic information management		50.0
TOTAL		114.0

Some economic consequences are not readily measurable, including the contribution that individuals educated in AMO science make when they switch fields, the ramifications of enhanced quality of communication and record-keeping permitted by microelectronics, the productivity gained because of the short recovery times from laser surgical procedures, and the savings in insurance and public assistance cost (not to mention the savings in human suffering) from, say, laser eye surgery to prevent blindness.

In this assessment of economic impact, no attention has been paid to the impact on the economy of measurement and measurement technology, an area that, as described in Chapter 2, is strongly dependent on AMO science. A 1967 study (R.D. Huntoon, "Concept of a National Measurement System," *Science* **158** (October 6), 67-71, 1967) found that approximately 6% of the GNP was generated by measurement-related activity. Value added to the output stream of manufactured products was estimated to be another 6%. Though this study was conducted 25 years ago, there is no reason to believe that industry is less dependent on measurement technology today.

As the global economy evolves, it is becoming increasingly apparent that the United States must rely on sophisticated products for its economic prosperity. The United States holds a competitive position in many high-technology areas, and those areas enabled by AMO science are generally those in which this country has excelled. The global economy is, however, intensely competitive, and being a world leader today does not ensure being an effective competitor tomorrow. The United States can retain a strong position in the global economy only if it maintains the infusion of new ideas and methods that research provides and can readily embody these ideas and methods in economically viable products. Clearly, future challenges will not be met by AMO (or any other) science alone, but progress in AMO science will be a central element in the configuration of our economic future.

6

International Perspectives in AMO Science

Given the importance of AMO science, it is not surprising that many nations have substantial research efforts in this area. Indeed, a recent meeting was organized by the United Kingdom's Science and Engineering Research Council to identify area of physics that would benefit from targeted funding. The four areas selected were real-time studies (sub-picosecond investigations of systems with ultrashort laser pulses), clusters, polarized electrons (which are powerful probes of atomic and surface processes), and ion-surface interactions—remarkably, each area falls in the domain of AMO science. Similarly, in Japan a number of priority areas for research have been identified, many of which involve AMO science—the atomic physics of multiply charged ions, activation of small inert molecules, theory of chemical reactions, carbon clusters, ultrafast and ultraparallel optoelectronics, and tunnelling characteristics of individual atoms.

As has been experienced in previous efforts to compare AMO science in the United States with that in other countries, the differences in infrastructures, funding methodologies, and political climates make the task of obtaining quantitative results most daunting. It is nevertheless possible to report opinions of the U.S. community and anecdotal statements of some of the leaders. Also, a citation analysis has been carried out and reported here, illustrating the relative health of U.S. AMO science as determined with that metric (see Appendix C).

In letter responses from leaders in the AMO field, it was an almost universal response that AMO science in the United States is comparable to or ahead of that in any other country, but that Germany and other Western European countries as well as Japan are moving forward rapidly, have closed gaps where they existed, and may soon move ahead. In some select areas, it was opined that the United

States already is lagging. It was especially emphasized that European Community countries and Germany in particular generally have more up-to-date and superior equipment and seem better supported. Statements like, "The last time I visited Germany, I felt like a poor relation," and "that position [of leadership] has suffered gradual and continual erosion, and we are at a point where many of the laboratories with which I am familiar are desperately short of funds to maintain and modernize their equipment," characterize many of the letter responses. Respondents to the FAMOS survey questionnaire also believe that the U.S. contribution to the total worldwide research and development effort in AMO science has decreased somewhat in the past 5 years (see Appendix D, Figures D.10a,b).

For critical technologies that are strongly impacted by AMO science, the picture is mixed. According to the U.S. Department of Commerce (DOC) Technology Administration (*Emerging Technologies: A Survey of Technical and Economic Opportunities*, Technology Administration, U.S. Department of Commerce, U.S. Government Printing Office, Washington, D.C., Spring 1990), Japan leads the United States in optoelectronics, advanced semiconductor devices, advanced materials, and high-density data storage. The United States leads Japan in sensor technology, medical devices and diagnostics, high-performance computing, and flexible computer-integrated manufacturing. The DOC report rates the United States ahead of Europe in all categories except digital imaging technology, a category in which the United States also lags Japan. The failure of U.S. industry to capitalize fully on the science and technology base of the country, which has been discussed widely, appears to extend to AMO science and technology as well.

The strength of the U.S. position in AMO science generally articulated by letter responses and questionnaire responses is demonstrated in an analysis of citations in the field, though the decline of leadership is not necessarily demonstrated. The citation analysis presented in Appendix C suggests that any such decrease is small. Little change is evident in the country of origin of papers citing work in AMO science over the past decade. Also, although the total number of citing papers with contributions from authors from the European Community and Japan is comparable to the number with contributions from U.S.-based authors, some 40% of the citing papers contain U.S. contributions, a figure that testifies to the strength of AMO science in the United States. The impact of the United States in AMO science was further illustrated by analyzing the origins of the most highly cited AMO papers published in 1989. Over 70% of these papers had contributions from U.S. authors. The United States is especially strong in research in rapidly evolving forefront areas. These were identified by analyzing the content of the most highly cited 1989 papers, and the examples chosen were laser cooling, diode laser development, femtosecond laser development, and C_{60}. On average, 56% of subsequent papers in these fields that cite the original 1989 work have contributions from U.S. authors, a number that is significantly higher than the 40% typical of AMO science as a whole. It is also interesting to note that while other countries may be particularly strong in some specific area, the United States makes a major contribution in all areas. The citation analysis suggests that the United States is very strong in emergent areas and has the flexibility and inventiveness to respond to new opportunities. The leadership role the United States plays in AMO science is also reflected in the responses to the questionnaire. On average, over 70% of respondents rated the U.S. research and development effort as strong/very strong relative to the rest of the world in the area of innovation. The ranking is not quite as high in terms of productivity, equipment, and facilities. Nonetheless, in each category significantly more respondents considered the U.S. position as strong/very strong than considered it weak/very weak.

AMO science is becoming increasingly international. Approximately 40% of respondents to the questionnaire indicated that they are involved in collaborative research and development programs with groups outside the United States. Of these, about one-half stated that they are very involved in such programs, and very few stated that they are only peripherally involved. The majority also indicated that their level of involvement has increased in the past 5 years. For highly cited papers, however, the citation analysis indicates that only about 13% of papers in AMO science contain contributions from authors in institutions in two or more countries.

APPENDIXES

APPENDIXES

A
Nobel Prizes Awarded in AMO Science Since 1964

1964

Nicolai Gennediyevich Basov

Nationality: Soviet
Area of concentration: Quantum electronics
Basov played an essential role in the invention of quantum microwave amplification devices (masers) and light amplifiers (lasers), which operate on the principle of stimulated emission of radiation. He collaborated with Aleksandr Prokhorov, with whom he shared the Nobel Prize, to produce the first Soviet maser and did pioneering work on the use of semiconductors in lasers.

Aleksandr Mikhailovich Prokhorov

Nationality: Soviet
Areas of concentration: Quantum radiophysics and quantum electronics
The independent research of Prokhorov and Nicolai Basov in the Soviet Union and Charles Townes in the United States on stimulated emission of radiation in the microwave and optical regions of the spectrum led to the development of masers and lasers.

Charles H. Townes

Nationality: American
Area of concentration: Quantum electronics
Townes's invention of the maser was a result of his investigation into the means of using stimulated emission of atoms for amplification of microwaves. An essential ingredient of Townes's discovery was the creation of an inverted population of atoms.

1966

Alfred Kastler

Nationality: French
Areas of concentration: Optical spectroscopy and Hertzian resonances
Kastler's discovery in 1950 of double resonance and his combining of this method in 1952 with the technique of optical pumping resulted in new knowledge of atomic structure and led to the development of masers and lasers between 1952 and 1958 by Townes in the United States and Prokhorov and Basov in the Soviet Union.

Robert S. Mulliken

Nationality: American
Area of concentration: Structural chemistry
Through the application of quantum mechanics, Mulliken developed the theory of molecular orbitals, which provided new insight into the structure of the chemical bond. He also studied molecular spectra and isotope separation.

1967

Ronald G.W. Norrish

Nationality: British
Areas of concentration: Photochemistry and reaction kinetics
Norrish contributed much to the maturation of the field of photochemistry and to the study of the kinetics of very fast chemical reactions. In the development of the technique of flash photolysis, he added immeasurably to the understanding of processes as diverse as polymerization and combustion. Laser spectroscopy

methods based on this phenomenon can be applied to numerous other fields of physics and chemistry.

George Porter

Nationality: British
Areas of concentration: Photochemistry and reaction kinetics
Porter developed and refined flash photolysis, which offers a means of measuring extremely fast chemical reactions. This technique has proved valuable in studying a wide variety of important reactions throughout chemistry.

1971

Dennis Gabor

Nationality: British
Areas of concentration: Electron optics and holography
Gabor was awarded the Nobel Prize for his discovery of the principles underlying the science of holography. While his fundamental studies in the optics of holography were completed in the late 1940s, Gabor was unable to realize the potential of his theoretical work until after the invention of the laser in 1960.

Gerhard Herzberg

Nationality: Canadian
Areas of concentration: Molecular spectroscopy and structure determination
Herzberg was a leader in molecular spectroscopy, a method of study that can identify molecules and provide precise information on their electronic structure and motions. He performed pioneering work with free radicals, the highly reactive molecular fragments that occur as intermediates in chemical reactions.

1976

William N. Lipscomb, Jr.

Nationality: American
Areas of concentration: Borane chemistry and X-ray crystallography
Lipscomb, through skillful experiments and exacting calculations, delineated and organized the chemistry of boron-hydrogen compounds (boranes). His

work on boranes is unique in its depth and scope and reveals new aspects of chemical bonding, molecular structure, and chemical reactivity that have general applicability.

1981

Nicolaas Bloembergen

Nationality: American
Areas of concentration: Optics and quantum electronics
Bloembergen's formulation of a general theory to explain the response of matter to intense laser light led to his development of the new field of nonlinear optical laser spectroscopy. Methods based on this phenomenon can be applied to numerous other fields of physics and chemistry

Kenichi Fukui

Nationality: Japanese
Areas of concentration: Electronic structure and organic reactions
Fukui discovered that of the many electronic orbitals involved in molecular structure, only those of the highest energy dominate the reaction. Fukui found that these frontier orbitals could account for many organic reactions not otherwise understood.

Roald Hoffmann

Nationality: American
Area of concentration: Electronic structure of compounds
Hoffmann recognized the importance of both the energy and the symmetry of electronic orbitals in chemical reactions. His development of the theory of orbital symmetry has become an exceedingly practical instrument for a wide variety of chemical syntheses.

Arthur L. Schawlow

Nationality: American
Areas of concentration: Optics and laser spectroscopy
Schawlow's discovery of new techniques in high-resolution laser spectroscopy opened a new era in atomic and nuclear physics by making it possible to study optical transitions with a resolution limited only by their natural line widths.

Kai M.B. Siegbahn

Nationality: Swedish
Area of concentration: Chemical physics
Siegbahn investigated and elucidated the binding energies of atomic electrons by dislodging the electrons with soft X-rays, developing a highly sensitive technique called electron spectroscopy for chemical analysis, or ESCA.

1986

Dudley R. Herschbach

Nationality: American
Area of concentration: Molecular reaction dynamics
Herschbach was one of a group of physical chemists who were the prime movers in the development of molecular beam machines. In part through his efforts, the molecular beam technique moved from early experiments using limited types of atoms to a modern technique capable of using any molecule. He was also instrumental in the development of the theoretical models to describe reactions.

Yuan T. Lee

Nationality: American
Areas of concentration: Molecular reaction dynamics and photochemistry
Lee helped revolutionize the field of molecular reaction dynamics through his construction of the first crossed molecular beams apparatus capable of detecting all molecules rather than select types. He conducted many experiments using these types of instruments and developed insight into the detailed mechanisms of bond formation and breaking in chemical reactions.

John C. Polanyi

Nationality: Canadian
Area of concentration: Molecular reaction dynamics
Polanyi developed the experimental method of infrared chemiluminescence, which allows chemists to look at the internal state distributions of product molecules in a chemical reaction. He performed a systematic study of the influence of potential energy surface features on the energy distributions in product molecules.

Gerd Binnig and Heinrich Rohrer

Nationality: G. Binnig, German; H. Rohrer, Swiss
Area of concentration: Surface science
Binnig and Rohrer designed and developed the scanning tunneling microscope. In doing so, they were the first to demonstrate vacuum tunneling. Scanning tunneling microscopy is a technique that is able to image a surface on a variety of length scales and is capable of resolving individual atoms and molecules. The position and orientation of adsorbed molecules can be determined and manipulated.

1989

Hans G. Dehmelt

Nationality: American
Area of concentration: Atomic spectroscopy
Dehmelt used ion-trap spectroscopy to study electrons and other charged particles. He was the first to observe a single electron in a trap, opening the door to precise measurements of key electron properties. Similar techniques allowed Dehmelt and his collaborators to observe a quantum jump in a single ion.

Wolfgang Paul

Nationality: German
Area of concentration: Atomic physics
Paul developed an electromagnetic trap capable of holding a small number of ions for long periods of time. The so-called Paul trap and its cousin, the Penning trap, play an important role in modern spectroscopy.

Norman F. Ramsey

Nationality: American
Area of concentration: Atomic physics
Ramsey developed a technique of imposing two separate, oscillating electromagnetic fields on an atomic beam to induce energy-level transitions that forms the basis of the cesium atomic clock. He also helped develop the hydrogen maser, which is useful as a secondary time standard.

1991

Richard R. Ernst

Nationality: Swiss
Area of concentration: Magnetic resonance imaging
Ernst improved nuclear magnetic resonance techniques, and his contributions paved the way for magnetic resonance imaging (MRI), a biomedical technique for depicting tissues deep within the body.

1992

Rudolph A. Marcus

Nationality: American
Area of concentration: Physical chemistry
Marcus is widely known for his theory of electron-transfer reactions. His work provided simple mathematical expressions for how energy of a molecular system is affected by changes in the structure of reacting molecules and their nearest neighbors. Electron transfer is a fundamental step in photosynthesis, metabolism, xerography, and chemical storage of electrical energy.

B

Impact of AMO Science

Throughout this report the contributions of AMO science to areas of national need and to other fields of science are highlighted, and a number of opportunities are underscored. In Figure B.1, the panel illustrates these contributions in the form of a matrix. Rows of the matrix represent the agglomerated research areas of AMO science as used in the questionnaire sent to the community and reproduced at the end of Appendix D. Columns represent various areas of national need. The degree of shading in the blocks indicates the perceived current level of impact. This level was assigned by the panel following consideration of the material in the body of the report and of inputs from the community, but relying primarily on the knowledge and judgment of the panel regarding the various elements of the matrix. Clearly, matrix elements are not independent of one another, but more complicated correlations cannot be shown in this simple presentation.

The broad extent to which AMO science serves the needs of the nation is graphically evident from Figure B.1. Particularly striking is the relationship of research in all areas of AMO science with energy and defense. The importance of AMO research to environmental problems is also clearly apparent, and the influence in the areas of industrial technology, materials processing, and manufacturing is remarkable.

Especially conspicuous in the matrix is the pervasive importance of laser spectroscopy and the physics of coherent light sources to literally all of the needs areas listed. Similarly impressive are the strong enabling aspects of atomic and molecular structure, optical interactions, and collisions, not only to a number of applied areas but also to other fields of science.

The matrix bespeaks the significance of AMO science as an essential player in the economic, technological, and scientific future of the nation.

FIGURE B.1 Overview of contribution of atomic, molecular, and optical science to areas of national need.

AMO RESEARCH AREAS
Atomic and Molecular Structure and Properties
Optical Interactions with Atoms and Molecules
Atomic and Molecular Collisions and Interactions
Interaction of Atoms and Molecules with Solids and Surfaces
Studies of Special Atoms and Molecules
Laser Spectroscopy
Nonlinear Optical Phenomena
Quantum Optics
Optical Interactions with Condensed Matter
Ultrafast Optics
Physics of Coherent Light Sources

AREAS OF NATIONAL NEED														
Technological Infrastructure								Other Sciences						
Industrial Technology, New Materials, Processing, and Manufacturing	Information Technology, High-Performance Computing, and Communications	Energy	Environment, Pollution, Global Change	Defense	Health and Medical Technology	Space Technology	Transportation	Nation's Measurement System	Discovery and Testing of Global Scientific Principles	Astrophysics	Plasma Science	Atmospheric Science	Condensed-Matter Physics	Space Science	Nuclear and Elementary Particle Physics, Biology, and Other Sciences

Key:
■ Provides major contributions to the area of national need.
▨ Essential to progress in the area of national need.
▢ Important for progress and viability of the area of national need.
□ Direct benefits to need area less evident at present time.

C
Citation Analysis

To explore the state of AMO science in the United States, a citation analysis was conducted using data provided by the Institute for Scientific Information (Philadelphia). Papers in AMO science were identified through the journals in which they appeared, and the data set comprised all the papers published in the journals listed in Table C.1. Recent trends were studied by analyzing the country of origin of the most highly cited (and thus presumably most significant) papers and by looking for changes in the distribution of the countries of origin of papers citing earlier work in AMO science. The country of origin of a cited or citing paper was determined from the author affiliations. In the small fraction of cases where a paper had contributions from authors affiliated with institutions in two (or more) countries, each of these countries was credited with a publication and/or citation. Self-citations could not be excluded.

The countries of origin of the most highly cited papers in AMO science published in 1989 are shown in Figure C.1. These data include papers published in the journals listed in Table C.1, together with *Applied Physics Letters, Nature*, and *Physical Review Letters*. Only a fraction of the papers published in these latter journals are in AMO science, and the title of each individual paper was checked to determine if it was in the field. Each paper forming the sample analyzed in Figure C.1 (a total of 381 papers) has been cited at least 15 times since publication. However, the distribution evident in Figure C.1 was little changed if the cutoff was raised to papers with at least 20 citations (a total of 167 papers) or 25 citations (a total of 85 papers). Approximately 70% of the highly cited papers included in Figure C.1 have contributions from U.S. authors. Because it is reasonable to assume that the number of citations a paper receives correlates with its importance to the field, this statistic speaks to the strength of AMO science in the United States, especially in universities. More than 80% of the papers included in the sample in Figure C.1 had contributions from university-based authors, whereas only about 30% had contributions from authors affiliated with either government or industrial laboratories. Of the papers included in Figure C.1, 13% represent international collaborations with authors from institutions in two or more countries.

TABLE C.1 Journals Included in the AMO Citation Analysis

Advances in Atomic, Molecular, and Optical Physics

Advances in Chemical Physics
Applied Optics
Chemical Physics
Chemical Physics Letters
IEEE Journal of Quantum Electronics
IEEE Photonics Technology Letters
Infrared Physics
International Journal of Mass Spectrometry and Ion Processes
Journal of Chemical Physics
Journal of Electronic Spectroscopy and Related Topics
Journal of Lightwave Technology
Journal of Luminescence
Journal of Modern Optics
Journal of Molecular Spectroscopy
Journal of Optics
Journal of Physical Chemistry
Journal of Physics B
Journal of Quantitative Spectroscopy and Radiative Transfer
Journal of the Optical Society of America A
Journal of the Optical Society of America B
Molecular Physics
Optica Acta
Optical and Quantum Electronics
Optics Communications
Optics Letters
Optik
Optika I Spektroskopiya
Physical Review A
Progress in Optics
Zeitschrift für Physik D

NOTE: *Applied Physics Letters, Nature,* and *Physical Review Letters* were included when analyzing the countries of origin of the most highly cited papers in AMO science published in 1989.

To ascertain how the U.S. contribution to the total worldwide effort in AMO science has changed in recent years, the countries of origin of papers citing AMO papers published in 1981 and 1989 in the journals listed in Table C.1 were examined. Figure C.2 shows the origins of papers that cited the 1981 papers during the period from 1981 to 1983, and the 1989 papers during the period from 1989 to 1991. In this interval the total number of citations increased by about 36%, reflecting, in part, the growth of activity in the field. The data in Figure C.2 are therefore normalized to the total number of citations in each period to allow better visualization of trends. The data suggest that the U.S. contribution to the total worldwide effort in AMO science is substantial and has changed little in the past decade. Europe and Japan, however, also have strong programs in AMO science.

To evaluate the ability of AMO science in the United States to respond to new opportunities and new ideas, the titles of the highly cited papers included in Figure C.1 were examined to identify "hot topics." These were laser cooling, diode laser development, quantum chemistry/collision dynamics, femtosecond laser development, application of femtosecond lasers, and C_{60}. Six representative papers in each area were then selected and the countries of origin of papers citing this work in the period from 1989 to 1991 examined. These are shown in Figure C.3, expressed as a percentage of the total number of citations in each area. Typically, about 56% of citations are associated with U.S.-based authors. This number is significantly higher than the approximately 40% typical of AMO science as a whole (see Figure C.2), indicating that the United States is especially strong in emergent new areas and has the resources and inventiveness to respond to new opportunities.

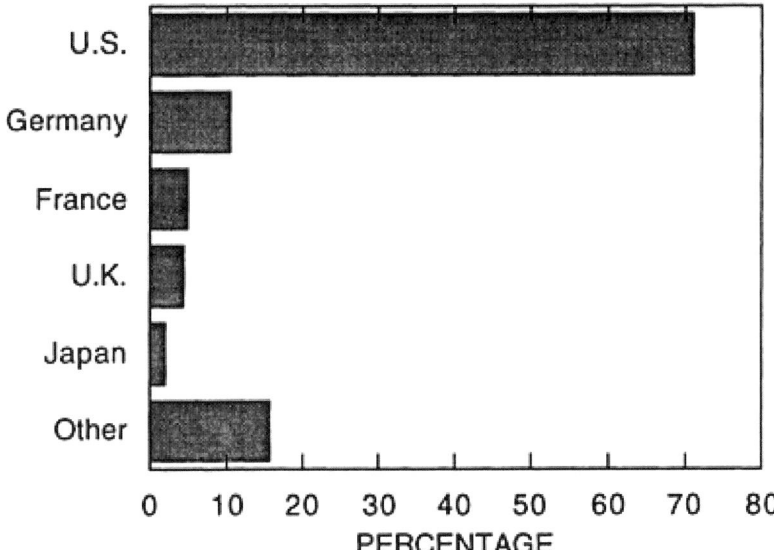

FIGURE C.1 Countries of origin of the most highly cited papers in AMO science published in 1989. The bars show the percentage of these papers with contributions from authors affiliated with institutions in the countries indicated.

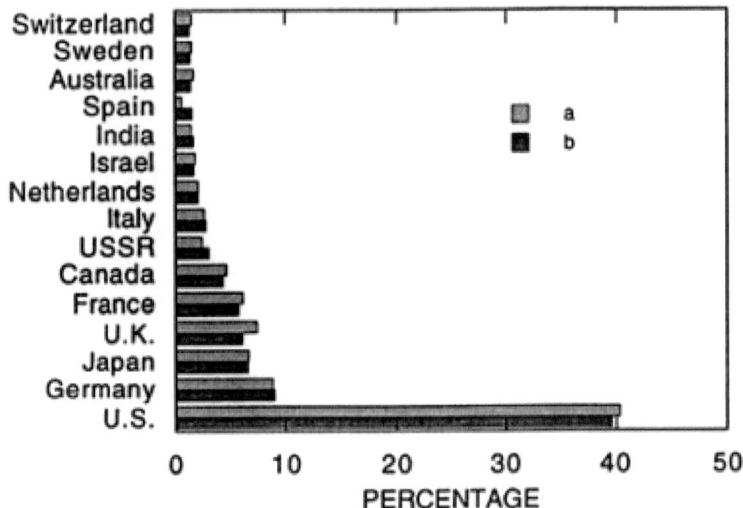

FIGURE C.2 Countries of origin of papers that cite papers in AMO science published in the journals listed in Table C.1 in (a) 1981 and (b) 1989. The citing intervals are 1981 to 1983 and 1989 to 1991, respectively. The horizontal bars represent the numbers of citations from each of the countries indicated expressed as a percentage of the total number of citations in each period.

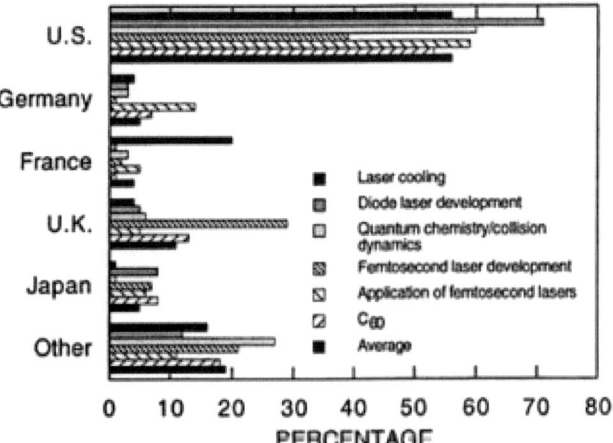

FIGURE C.3 Countries of origin of papers that cite papers in "hot topic" areas published in 1989, in the interval from 1989 to 1991. The horizontal bars represent the numbers of citations from each of the countries indicated expressed as a percentage of the total number of citations in each area.

D

Survey of AMO Scientists

As part of the FAMOS panel's assessment, a survey of AMO scientists was conducted by mail, using the questionnaire reproduced at the end of this appendix. The mailing comprised a letter of introduction, the five-page questionnaire and postage-paid return envelope, a Research Specialties Directory (also reproduced at the end of this appendix), and a postage-paid quick-response postcard. This was sent to members of the American Chemical Society, the American Physical Society (APS), the Institute of Electrical and Electronics Engineers, the Materials Research Society, the Optical Society of America, and the Society of Photographic and Instrumentation Engineers who expressed interest in areas related to the present assessment as defined by the division, topical groups, technical interest groups, and so on, to which each belonged. The mailing lists from the different societies were cross-referenced and culled to minimize duplicate mailings to the same person. Budgetary constraints required that the mailing be limited principally to members with PhD degrees.

A total of 19,650 questionnaires were mailed, of which 3,424 were returned completed. Approximately 2,000 recipients responded by stating that they would not complete the questionnaire because they were not practicing AMO scientists. The overall response rate of approximately 28% is typical for surveys of this type. The number of questionnaires obtained, however, is sufficient to ensure a diversity of opinion and circumstance and allows the data to be broken down and analyzed by subgroups. Nonetheless, the possibility exists that the results of the survey will be biased by nonresponse; that is, the opinions of those AMO scientists who did not respond are not represented. It is difficult to estimate what fraction of practicing AMO doctoral scientists actually received and returned a questionnaire. The categories included in setting up the mailing list were purposely kept broad to ensure good coverage of the field, and it was expected that the list would include many non-AMO scientists who might not respond at all. The 1990 APS employment survey reported 2,725 doctoral scientists actively employed in AMO physics in 1989, and that this number had remained essentially constant for several years. The panel's survey indicates that ~41% of AMO scientists hold degrees in physics, suggesting that there are about 6,000 to 7,000 AMO doctoral scientists active today. The completed questionnaires received therefore appear to represent a sampling of the opinions of approximately one-half of practicing AMO scientists.

EDUCATION AND EMPLOYMENT PROFILE

As expected, given the selection criteria applied at the time of mailing, the majority of respondents (93%) had a PhD degree. Forty-six percent indicated that they were employed in universities/colleges, 30% in industry, and 18% in federally funded research and development (R&D) laboratories or government (to be termed simply government). Sixteen percent described their R&D activities as principally atomic, 46% as principally molecular, and 38% as principally optical. Approximately equal numbers of respondents characterized their work as being principally basic or principally applied. Eighty percent described their R&D activities as principally experimental. Of the 20% whose activities were principally theoretical, 76% considered their work computationally intensive.

The year that respondents received their highest degree is shown in Figure D.1, broken down by employment sector and principal area of research. The overall distributions reflect the effects of retirement and illustrate disproportionate growth in optical science over the past 30 years. A similar growth in industrial employment is also noted. The distribution of specialties in which respondents working in industry, government, and universities received their highest degrees is shown in Figure D.2. A total of 41% of the respondents received their highest degree in some area of physics, 36% in chemistry, and 23% in engineering or some other field. Respondents with degrees in optical physics, applied physics, electrical engineering, and materials science had the highest probability of being employed in industry. The places in which respondents work is shown in Figure D.3a, broken down by principal area of research. The difference in emphasis between work conducted in industry and in universities is clearly illustrated. In industry the focus is on optical science; in universities molecular science is dominant.

The distribution of times for which respondents have been employed by their present employer is presented in Figure D.3b. It is interesting to note that mobility appears highest in the industrial sector and that although some 38% of respondents have changed employers in the past 5 years, only ~18% received their highest degree in this interval.

The principal job positions of respondents employed in industry, government, and academe are shown in Figure D.4. In total, only 5% of respondents indicated that they held temporary positions. The professional societies to which respondents belong are indicated in Figure D.5. The number of different societies included in Figure D.5 illustrates the breadth of AMO science and the range of interests of its practitioners. On average, each respondent is a member of 2.3 societies.

Figure D.6 illustrates how the breakdowns between atomic, molecular, and optical, basic and applied, and experimental and theoretical R&D activities depend on employment sector. In industry, 87% of respondents are involved in applied work, whereas in universities 75% undertake basic research. The relative emphasis on basic and applied work is about equal in government laboratories. Theoretical activity is greatest in universities and significantly less in industry.

RESEARCH AND DEVELOPMENT ACTIVITIES AND TRENDS

The research specialties of respondents are shown in Figure D.7, broken down by employment sector. The emphasis on optical science in industry and on molecular (and atomic) science in universities is again evident. The enabling aspect of AMO science is also demonstrated in that a large majority of respondents in industry noted that their work interfaced with other areas of science and technology.

A substantial fraction of respondents (66% in industry, 77% in government, and 73% in universities) indicated that they are involved in collaborative programs with other groups in the United States. A smaller but still sizable fraction (32% in industry, 41% in government, and 45% in universities) reported collaborations with groups outside the United States. Of those respondents involved in collaborative research, 64% in industry, 62% in government, and 55% in universities described their work as cross-disciplinary, which further emphasizes the breadth of AMO science and its many applications. Their level of involvement in collaborative programs is illustrated by Figure D.8. Approximately 50% described themselves as being very involved, whereas only ~10% showed their involvement as peripheral. The large majority, over 85%, indicated that in the past 5 years their level of involvement has remained the same or increased; indeed, ~30% indicated that their involvement has increased substantially. This may reflect the broadening impact of the field and/or changes occurring in response to funding pressures.

The utilization of user facilities by respondents is shown in Figure D.9a. Only ~20% indicated that more than 50% of their experimental or computational work is undertaken at user facilities, whereas approximately 50% do not utilize such facilities at all. The utilization patterns change little between employment sectors (or major research area). A weighted average of the responses suggests that some 22% of the work in AMO science is undertaken at user facilities. The responses also suggest that computational theorists are somewhat more likely to use user facilities than are experimenters. The data indicate that ~29% of computational work is undertaken at user facilities. Figure D.9b shows how the respondents' utilization of user facilities has changed in the past 5 years. No strong trends are evident. Approximately 17% indicated their activities at user facilities have increased, whereas 9% said these have decreased.

Figure D.10a shows the countries that respondents listed as having significant R&D activities in their major area. It is apparent that only a few respondents work in areas in which there is no competition from outside the United States. Competition from Japan is especially strong in optical science and from Germany in atomic and molecular science. Figure D.10b indicates how respondents believe the U.S. contribution to the total worldwide effort in their major area has changed in the past 5 years. Clearly, the perception is that in all areas of AMO science the U.S. contribution has decreased somewhat in this interval.

Figure D.11 shows how respondents rated R&D efforts in their area in the United States in comparison to those in other countries in terms of innovation, productivity, equipment, and facilities. The U.S. effort ranks well in each area, especially in innovation, where ~72% of respondents rated U.S. performance as strong/very strong, that is, a rating of 1 or 2. The rating in terms of productivity, equipment, and facilities is somewhat lower, but 48%, 50%, and 45% of respondents, respectively, still believe that the U.S. position is strong/very strong.

How respondents regard the balance between theory and experiment is illustrated in Figure D.12. It is apparent that researchers in all three employment sectors, and in each major research area, believe the balance between theory and experiment to be about right. To explore this question further, the data were sorted according to whether the respondents' interests were principally experimental or principally theoretical, but as is evident from Figure D.12, even theorists feel that the present balance between theory and experiment is reasonable.

The sources of R&D funding for the groups or teams to which respondents belong are shown in Figure D.13. A large percentage (~80%) of the university research groups represented receive support from federal agencies, highlighting the importance of federal funding in maintaining a strong university research effort. However, a significant number of university groups receive funding from other sources. In particular, ~19% and 13% receive support from industrial contracts and corporate donations, respectively (the largest number of these going to support optical and molecular science). Approximately 82% of industrial groups represented in the survey receive internal support, although ~35% also obtain funding from federal agencies. As expected, federal monies support the bulk of the research effort in government laboratories.

The total amounts and sources of annual support for AMO science listed by respondents are presented in Figure D.14a, which shows an aggregate annual support of $610M. This figure, however, does not include industrial funding from internal sources. Approximately 71% of university respondents served as principal investigator or co-principal investigator on one (or more) grant, contract, or donation. This fraction was somewhat smaller in government (~60%) and much less in industry (~29%). The largest individual source of support is DOD, with an annual support level of $204M, followed by DOE ($103M), NSF ($59M), and NASA ($52M). Other federal support totals $48M. Industrial contracts ($69M) are also a significant source of funding. Private foundations, corporate donations, and state funds contribute ~$48M. Given that not all AMO scientists responded to the questionnaire, these numbers represent lower bounds to the annual funding level for AMO science, which is likely to be ~$1B. Nonetheless, it is clear that DOD is a major supporter of AMO science and that changes in its funding patterns could have a major impact on the field, especially in the industrial sector, where DOD support is dominant. In contrast, the present survey indicates that NSF is the largest supporter of university-based AMO R&D (~25%, $55M), followed by DOD (~21%, $47M), and DOE (~12%, $26M). Thus the fate of university-based AMO research is strongly tied to that of NSF. The total federal support listed by university-based researchers is $165M.

The mean and median values of the annual support that individual respondents receive from NSF, DOE, DOD, and NASA are shown in Figure D.14b . These may not reflect the true median/average grant sizes for those agencies, however, because a given respondent may have combined support from more than one grant/contract from a given agency or may be co-principal investigator and thus may have reported only his/her share of the total. It is apparent that the mean/median support levels provided to individual investigators by NSF and NASA are significantly lower than those provided by DOD and DOE. The large differences between the median and the mean values in the case of DOE, DOD, and NASA result because these agencies support a number of multimillion dollar projects that significantly increase the mean.

Approximately 76% of respondents receiving direct research support indicated that they were involved in mission-oriented research. The distribution of responses among the mission categories (energy, health and the environment, space science, defense, and commerce and technology) is shown in Figure D.15. R&D related to defense dominates in industry and is important in university and government laboratories, where, however, it is almost equaled by energy-related R&D. There are also significant activities in the areas of health and the environment, and space science. In optical science, defense dominates mission-oriented R&D. For atomic and molecular science, energy is the most important category, with defense a close second. AMO science is, however, clearly important in all mission areas.

The percentage of the total R&D funding that respondents receive that is used in salary/stipend support is shown in Figure D.16. Clearly, such support accounts for a major fraction of the research budget in each sector, leaving little flexibility, especially as regards purchase of major new research instrumentation. This inability to upgrade equipment can limit R&D progress and output and reduces the exposure of graduate students to state-of-the-art equipment and techniques, a vital part of their training.

The distribution of the number of different sources from which individual respondents with R&D funding receive that funding is shown in Figure D.17a. Approximately 60% indicated that they receive support from only one or two sources. Fewer than 10% obtain funding from more than five sources. Figure D.17b shows how the number of sources from which respondents receive funding has changed in the past 5 years. This number appears to have remained largely unchanged in recent years.

The change in respondents' R&D support in the past 5 years is shown in Figure D.18. Approximately 51% indicated that in real terms their support has decreased, and only ~25% that it has increased. The situation appears to be worst in government laboratories. These figures may also underestimate the real decrease in funding, because respondents who had no current R&D funding but may have had support within the past 5 years were instructed not to answer this question. No significant differences in the funding history for each specialty area, that is, atomic, molecular, and optical, were evident. As demonstrated in Figure D.19, a large majority, ~71%, of respondents believe that funding for strong programs should not be increased at the expense of the total number of programs supported.

The percentage of time that respondents spend pursuing R&D support is shown in Figure D.20. Although ~61% of respondents reported spending less than 20% of their time seeking funding, a significant number, ~12%, stated that they spend greater than 50% of their time in this pursuit. However, researchers in universities tend to spend a somewhat larger average fraction of their time seeking support than do their colleagues in industrial or government laboratories.

The averaged relative priorities assigned to different categories are shown in Figure D.21, broken down by employment sector, and exhibit a number of prejudices. Respondents in all sectors, but especially universities, assign high priority to support of single-investigator, small-scale programs, which are the very heart of AMO science. Similarly broad support exists for funding young investigators, graduate students, and postdoctoral/research associates. It is also agreed that capital equipment items deserve strong support. There is unanimity that new user facilities, advanced computer facilities, centers of excellence, and soft-money positions are not of the highest priority. Respondents in government and industrial laboratories feel that existing user facilities and consortia/multi-investigator programs/interdisciplinary research deserve quite high priority, but workers in universities are somewhat less enthusiastic.

HUMAN RESOURCES AND DEMOGRAPHICS

The size of R&D groups or teams to which respondents belong is shown in Figure D.22. As might be expected, interaction with graduate and undergraduate students is greatest for respondents at universities/colleges. Indeed, ~67% of such respondents reported that they interact with one or more undergraduate research students, and 79% indicated that their research group contains one or more graduate students. Approximately 27% of respondents from government indicated interaction with undergraduate students, and ~41% involvement with graduate students. The corresponding numbers in industry are much lower, ~13% and ~18%, respectively. University and government laboratories also provide continued training for postdoctoral/research associates; ~56% and ~65% of respondents, respectively, in these sectors noted that they interact with one or more such researchers. In contrast, only ~23% of workers in industrial laboratories reported such interactions. A large fraction of the respondents in all these employment sectors reported interactions with one or more other research staff/faculty, although this is particularly high in government laboratories. Technical support is strongest in industrial and government laboratories, where somewhat more than 70% of respondents interact with one or more technicians. This figure is substantially smaller in universities (~35%). Clerical and administrative support in all three sectors appears comparable.

The fractions of respondents reporting changes in the major emphasis of their R&D activities in the past 5 years is shown in Figure D.23. Only ~45% of university researchers have changed their major R&D emphasis in this interval, which contrasts with the ~66% of industrial researchers who did. The major reasons for change appear to be changes in institutional direction/emphasis (except in universities), identification of new problems, funding and employment opportunities, and the availability of new technologies. Taken together, funding difficulties, employment difficulties, and the availability of new computer or user facilities do not appear to be major driving factors for changes in emphasis. These responses indicate that the evolution of the field is being governed principally by new scientific opportunities.

Information regarding hiring in the past 3 years to fill temporary positions is presented in Figure D.24. These data may be biased in that multiple returns from a large research group could result in an overestimation of available positions. However, given an overall response rate of only ~50%, and the typically small size of AMO R&D groups or teams, this effect should be relatively small. Also, comparisons of past and future trends are not affected by this bias. The data suggest that a majority (~60%) of R&D groups in government laboratories have hired, or tried to hire, AMO scientists to fill temporary positions. This figure is somewhat less (~46%) in universities. In industrial laboratories the situation is different, with only ~23% of such laboratories having attempted to fill temporary positions. Typically, in each sector groups that have sought to fill temporary positions have had only one, two, or possibly three openings. These positions appear to arise equally from refilling a temporary position and from creation of new positions.

Anticipated recruiting patterns for temporary positions for the next 3 years are illustrated in Figure D.25a. A large fraction (~70%) of researchers in industrial laboratories do not anticipate any new hires in the next 3 years. The situation is somewhat better in government and university laboratories, where ~60% and 50% of respondents, respectively, anticipate new openings. Again those groups that anticipate having positions to fill expect typically only one or two openings. Comparison of Figure D.24 and D.25a provides little optimism that the availability of temporary positions will increase in the near future. Taken together, respondents in government, industry, and universities estimate there will be 740, 540, and 1,330 temporary positions, respectively, in these sectors in the next 3 years; that is, respondents expect an average of 870 temporary positions in AMO science per year. This availability is slightly lower than that reported (~930 per year) for temporary positions in the past 3 years. Figure D.25b shows the distribution of specialties in which recruitment is considered likely. These distributions vary according to employment sector; industry emphasizes interfaces with other areas of science and technology, and optical sciences in general. Government laboratories place emphasis on interface areas, but areas such as atomic and molecular collisions and structure, and their optical interactions and spectroscopy, are also considered important. Universities place emphasis on these latter areas.

As illustrated in Figure D.26, somewhat over one-half the respondents from each employment sector indicated that their research group or team hired, or tried to hire, new faculty/staff to long-term or permanent positions in the past 3 years. Few respondents, however, reported more than two openings. In universities, only ~13% of openings are associated with failure to get tenure, the remainder arising approximately equally from retirement and creation of new positions. In government and industrial laboratories the majority of openings resulted from the creation of new positions.

Anticipated recruitment patterns for permanent or long-term positions are shown in Figures D.27a-d. Less than one-half the respondents in each employment sector anticipate hiring within the next 3 years, and comparison with Figure D.26 suggests that the job market in the immediate future will be tougher than in the past. Indeed, respondents in government, industry, and universities expect the availability of 360, 980, and 910 long-term positions, respectively, in these sectors. This corresponds to an average of ~750 positions per year, significantly less (~30%) than the average of ~1,070 positions per year reported for the past 3 years. Researchers in government laboratories are most pessimistic about the future, perhaps a response to uncertainties concerning the future role of government laboratories. The distribution of specialties in which recruiting is considered likely is shown in Figure D.27b. The emphasis in industry is on applications and optical science, and that in government and universities is more on atomic and molecular collisions and structure and their optical interactions and spectroscopy. As indicated in Figure D.27c, the majority of openings in universities (~51%) will be nontenured faculty positions, with ~31% tenured positions.

Only ~11% will be for research faculty/research scientist positions. In industry and government, most openings will be for members of technical staff or research scientists.

The likely R&D interests of the persons to be recruited are shown in Figure D.27d. The emphasis in industrial hiring will be on people with a background in optical science, and, to a lesser degree, molecular science. A similar bias is evident in the case of government laboratories. Molecular science will lead recruitment efforts at universities, which are more evenly divided among the three AMO categories. As expected, emphasis in industry will be on applied science, and in universities on basic science. In each employment sector, fewer than 16% of anticipated openings will be in theory, corresponding to a total of ~110 positions per year. Of these theorists, it is expected that ~80% will work on computationally intensive problems.

As indicated in Figure D.28, a substantial majority of respondents, especially those in industry, believe that human resources in AMO science are adequate to meet existing and future needs in terms of both quantity and quality. As demonstrated in Figure D.29, respondents do not believe that the importance of AMO science is decreasing. Indeed, researchers in universities believe that, at least in this setting, its importance is increasing significantly.

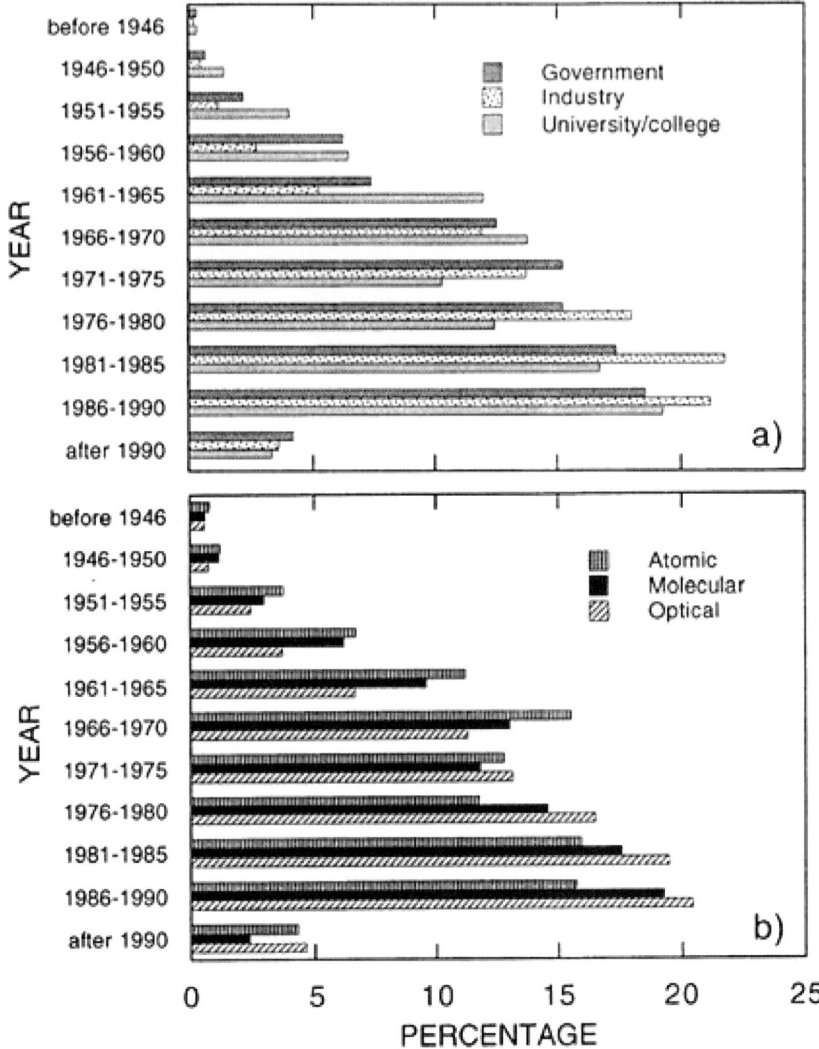

FIGURE D.1 The year respondents received their highest degree, broken down by (a) employment sector and (b) principal area of research. The bars represent the percentage of respondents in each sector or area who obtained their highest degree in the time intervals indicated.

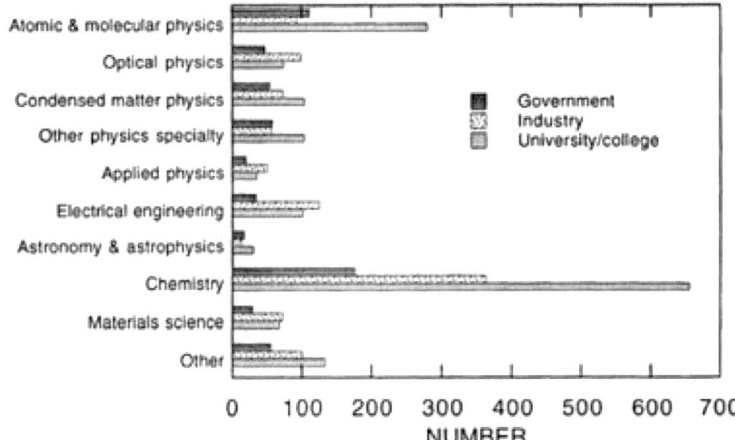

FIGURE D.2 The specialty in which respondents obtained their highest degree. The bars represent the number of respondents in each employment sector who obtained their degree in the specialties indicated.

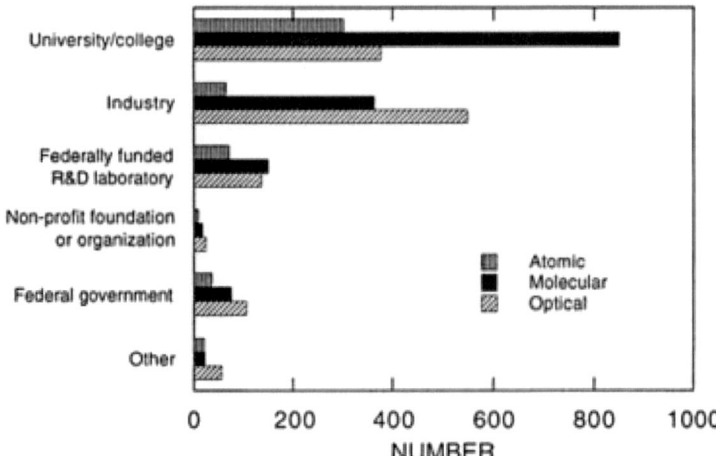

FIGURE D.3a Employment characteristics of respondents, broken down by major research area. The bars represent the number of respondents in each area in the employment sectors indicated.

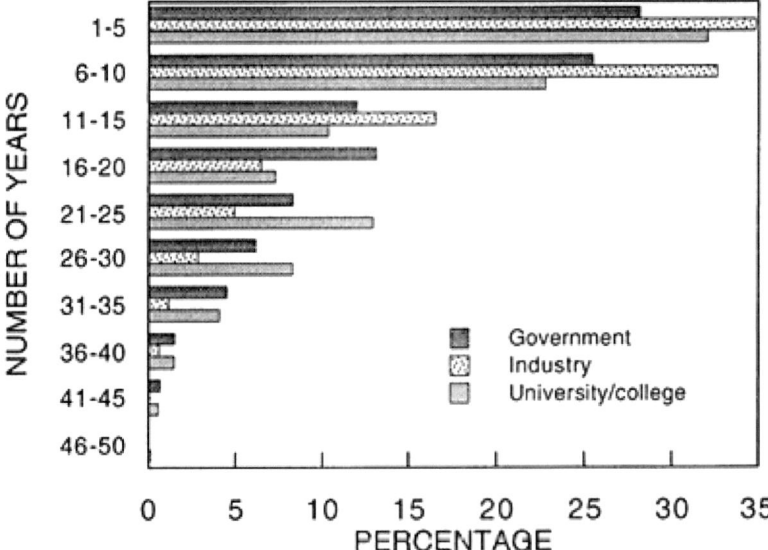

FIGURE D.3b The distribution of number of years that respondents have been employed by their present employer. The bars represent the percentage of respondents in each employment sector who have been employed for the time intervals indicated.

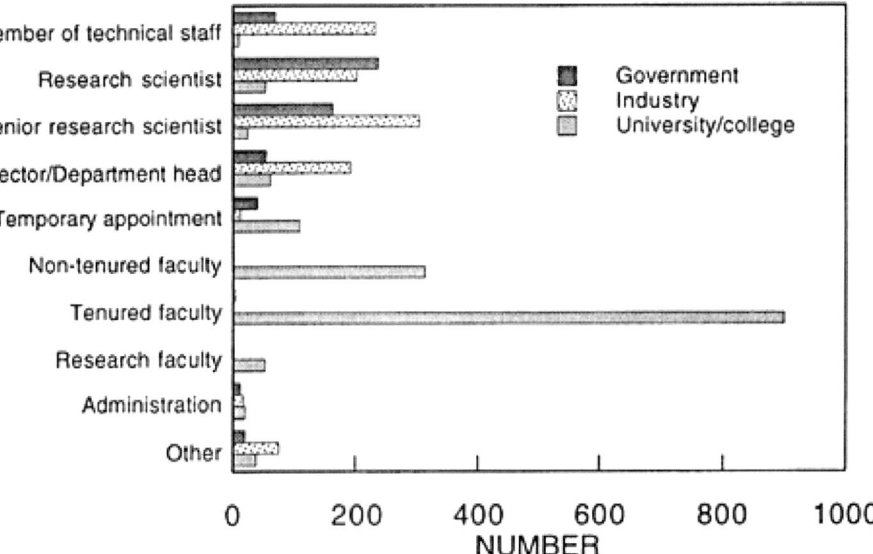

FIGURE D.4 Principal positions of respondents, broken down by employment sector. The bars represent the number of respondents in each sector who hold the positions indicated.

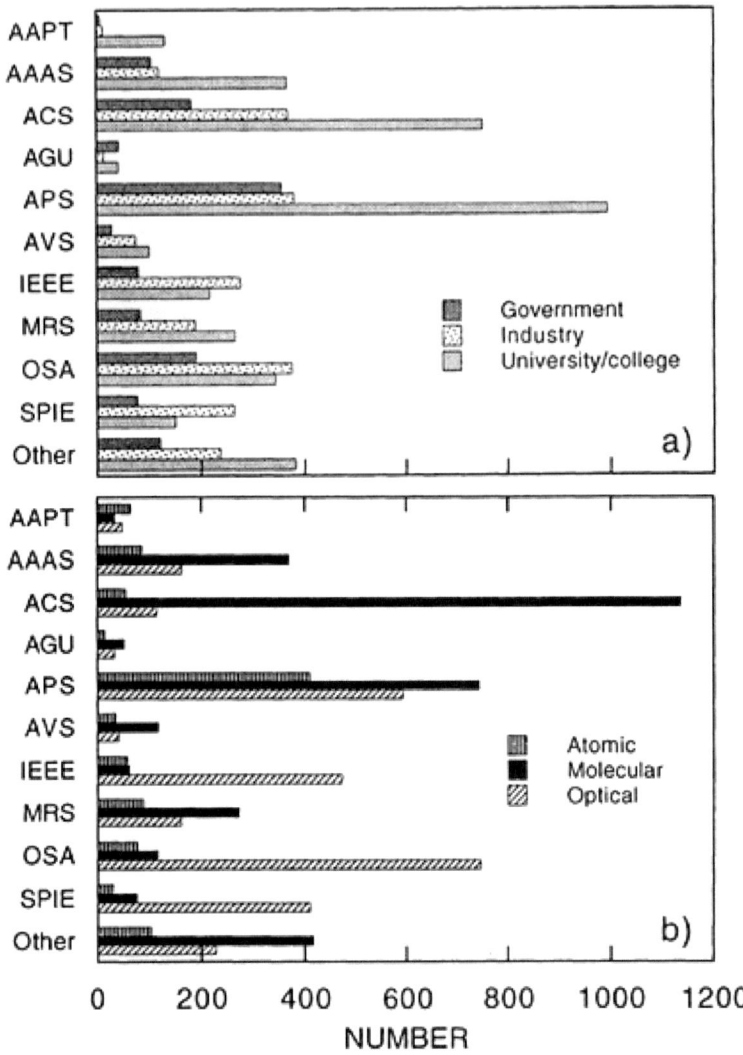

FIGURE D.5 Professional societies to which respondents belong, broken down by (a) employment sector and (b) major research area. The bars represent the number of respondents in each sector or area who are members of the societies indicated. [American Association of Physics Teachers (AAPT), American Association for the Advancement of Science (AAAS), American Chemical Society (ACS), American Geophysical Union (AGU), American Physical Society (APS), American Vacuum Society (AVS), Institute of Electrical and Electronics Engineers (IEEE), Materials Research Society (MRS), Optical Society of America (OSA), Society of Photographic and Instrumentation Engineers (SPIE).]

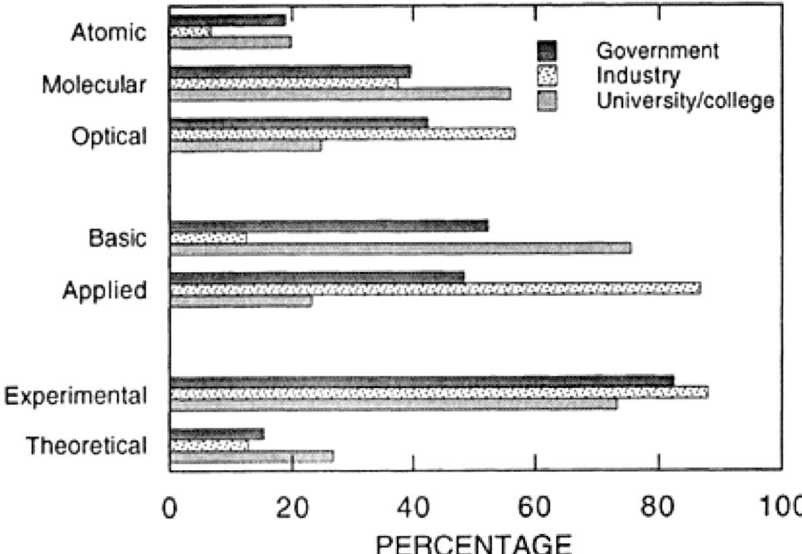

FIGURE D.6 Principal R&D activities and interests of respondents. The bars represent the percentage of respondents in each employment sector who characterized their principal R&D activities as indicated.

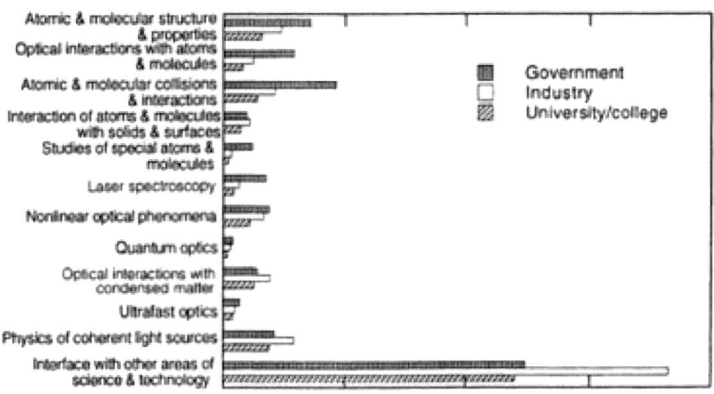

FIGURE D.7 Research specialties of respondents, broken down by employment sector. The topics included in each specialty are shown in the Research Specialties Directory (reproduced at the end of this appendix). The bars represent the number of responses in each listed specialty normalized to the number of respondents in each sector.

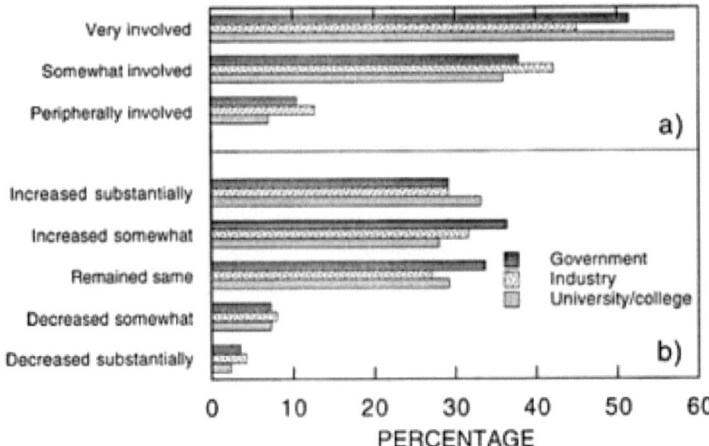

FIGURE D.8 (a) Level of involvement of respondents in collaborative research programs and (b) the history of this involvement, broken down by employment sector. The data set is limited to respondents undertaking collaborative R&D. The bars represent the percentage of these respondents whose level of involvement, and change in involvement over the past 5 years, is as indicated.

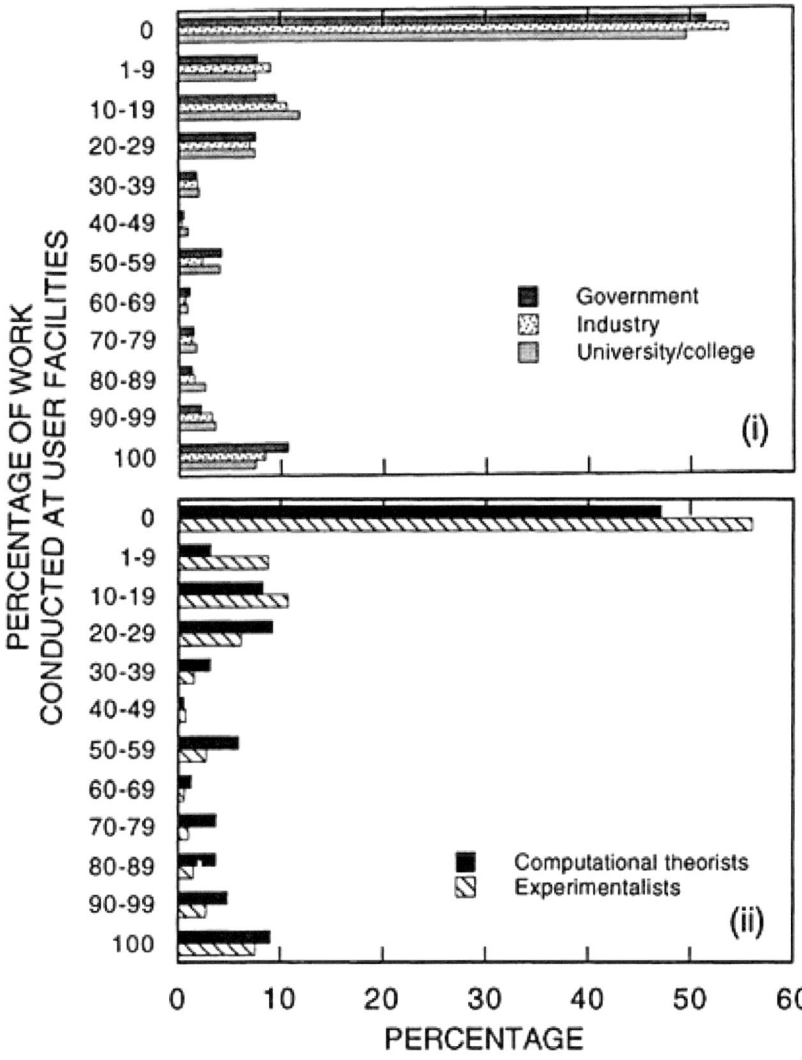

FIGURE D.9a Utilization of user facilities, broken down by (i) employment sector and (ii) experimentalists and computational theorists. The bars show the percentage of respondents in each sector or activity who characterized their usage as indicated.

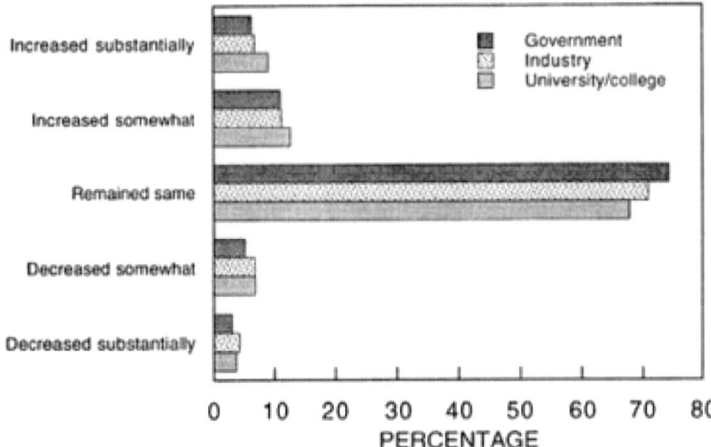

FIGURE D.9b Change in percentage of work conducted at user facilities over the past 5 years. The bars show the percentage of respondents in each employment sector who characterized the change in their involvement as indicated.

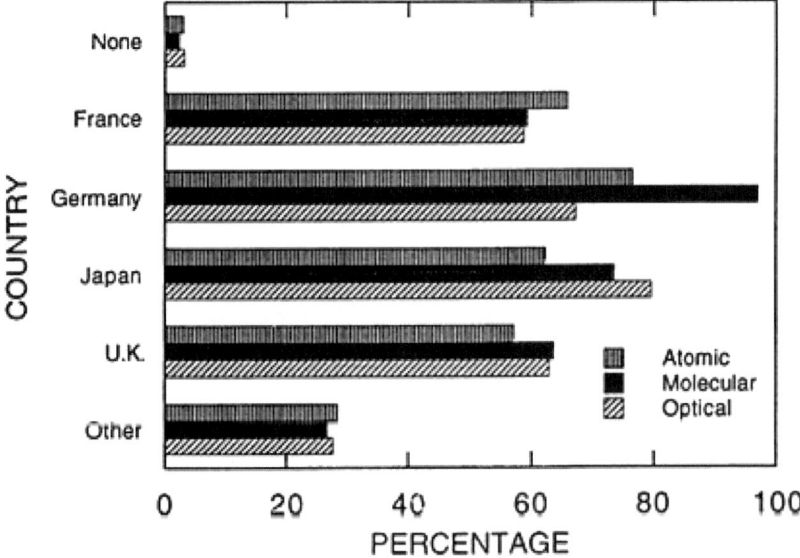

FIGURE D.10a Countries having significant R&D in the respondents' major area, broken down by major research area. The bars represent the percentage of respondents in each area who listed the countries indicated.

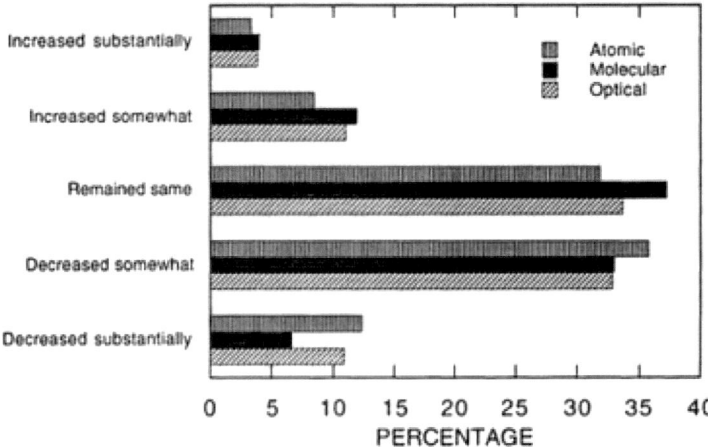

FIGURE D.10b Change in the relative U.S. contribution to the total world effort in the respondents' major research area over the past 5 years. The bars represent the percentage of respondents in each major research area who characterized the changes indicated.

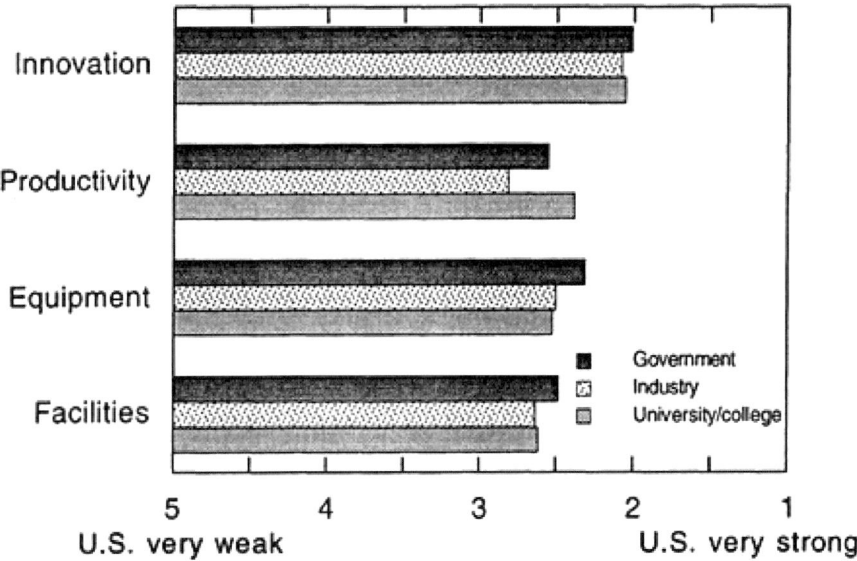

FIGURE D.11 Ranking of U.S. research efforts in the respondents' major research area in comparison to those in other countries, broken down by employment sector. The bars represent the mean ranking for each category indicated on a scale of 1 to 5.

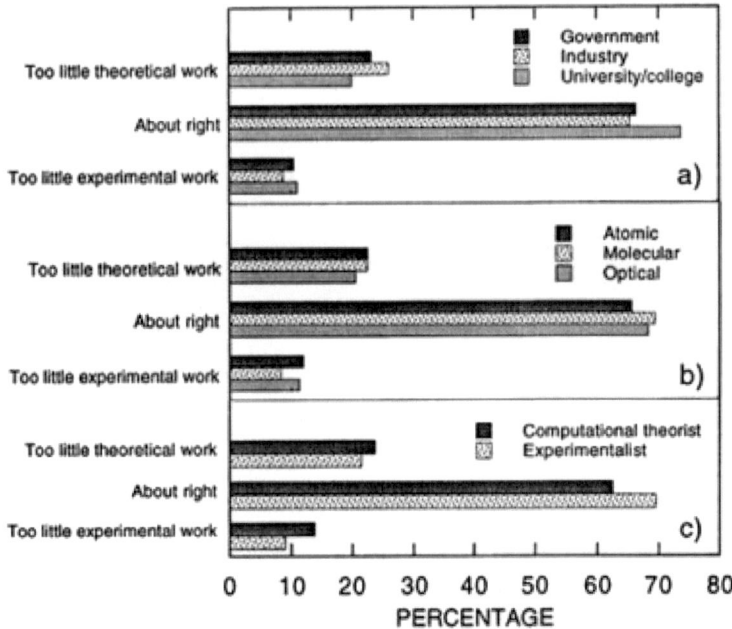

FIGURE D.12 Balance between theory and experiment, broken down by (a) employment sector, (b) major research area, and (c) experimentalists and theorists. The bars show the percentage of respondents who characterized the balance as indicated.

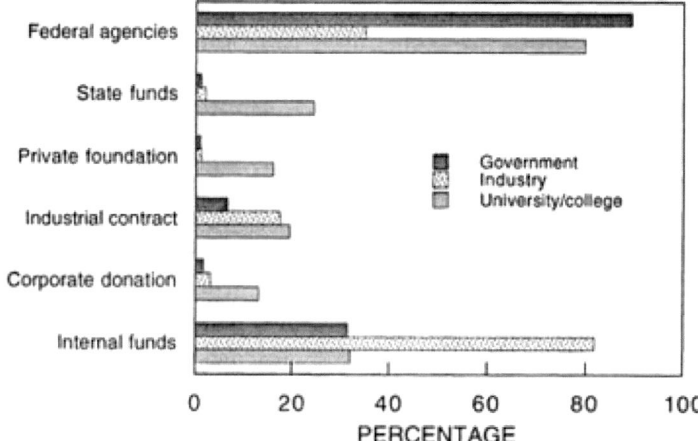

FIGURE D.13 Sources of funding for the groups or teams to which respondents belong. The bars represent the percentage of respondents in each employment sector who reported support from each indicated source.

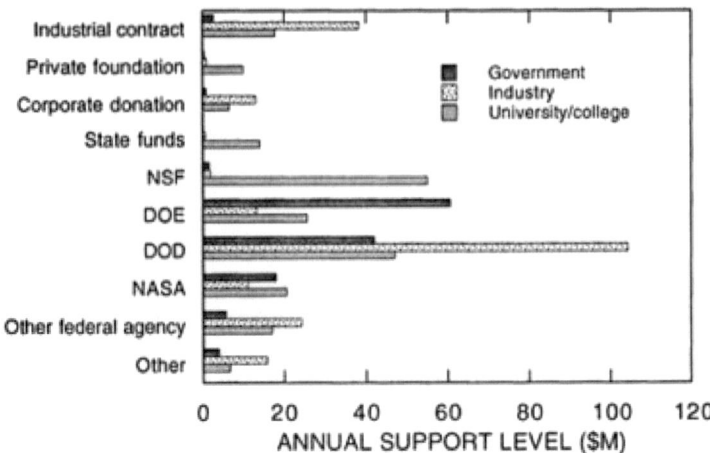

FIGURE D.14a Annual support levels for AMO science reported by respondents, broken down by employment sector. The bars represent the total funding from each source.

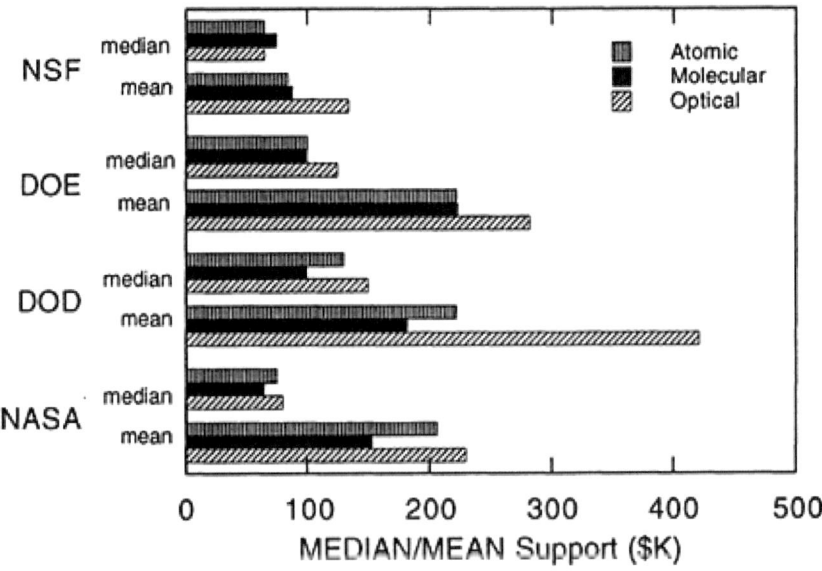

FIGURE D.14b The mean and median values of annual support provided by NSF, DOE, DOD, and NASA.

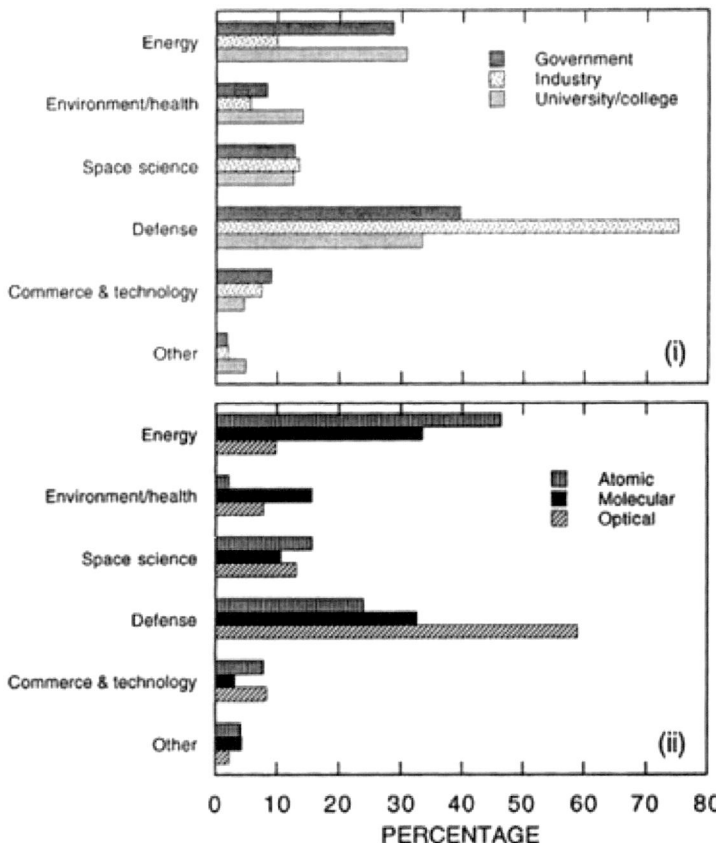

FIGURE D.15 Involvement of respondents in mission-oriented research, broken down by (i) employment sector and (ii) major research area. The bars represent the percentage of respondents with direct research support who reported undertaking research in each indicated mission area.

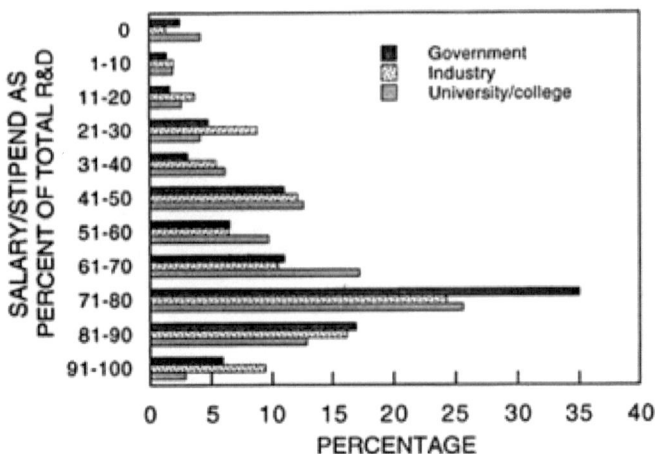

FIGURE D.16 Percentage of total R&D funding used by respondents in salary/stipend support, broken down by employment sector. The bars represent the percentage of respondents with direct research funding who indicated each listed support level.

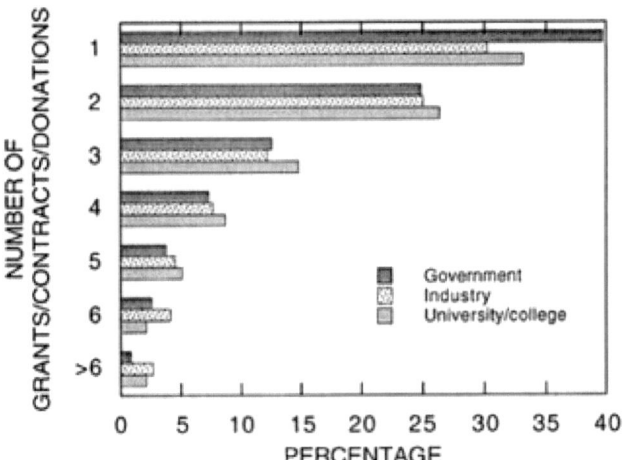

FIGURE D.17a Number of sources from which respondents receive research support, broken down by employment sector. The bars represent the percentage of respondents with direct research funding who reported each of the indicated number of sources.

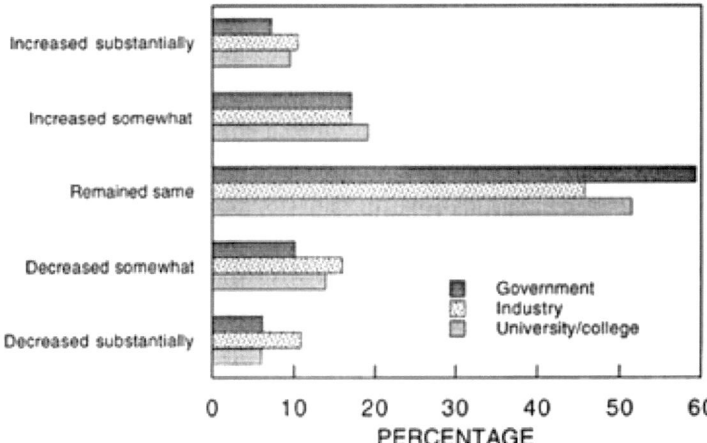

FIGURE D.17b Changes in the number of funding sources over the past 5 years, broken down by employment sector. The bars represent the percentage of respondents with direct research funding who characterized the change in number as indicated.

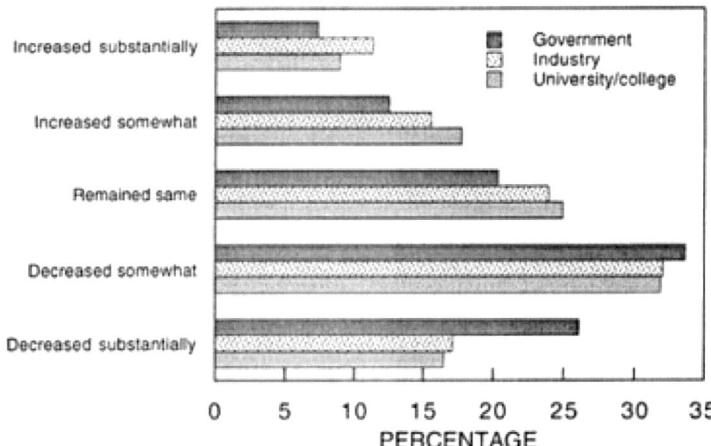

FIGURE D.18 Change in total research support over the past 5 years, broken down by employment sector. The bars represent the percentage of respondents with direct research funding who characterized their change in support as indicated.

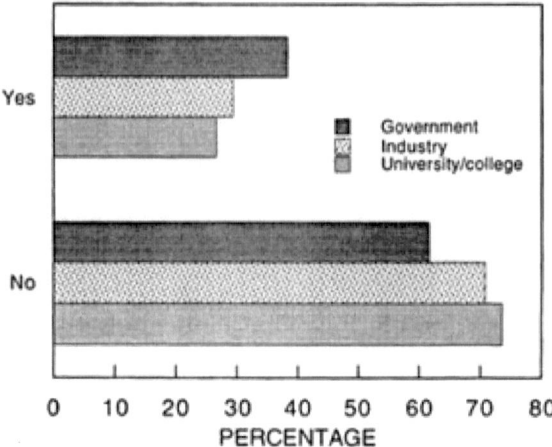

FIGURE D.19 Future funding patterns. The bars represent the percentage of respondents in each employment sector who r esponded "yes" and "no" to the question, Should support for the strongest programs be increased at the expense of the total number of programs supported?

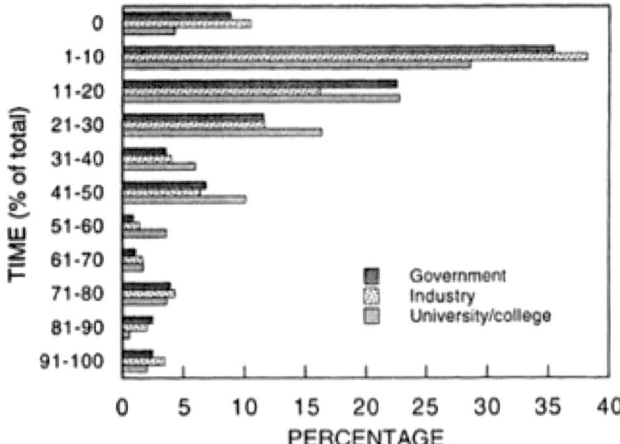

FIGURE D.20 Percentage of time that respondents spend pursuing research support. The bars represent the percentages of respondents in each employment sector who reported the time commitments shown.

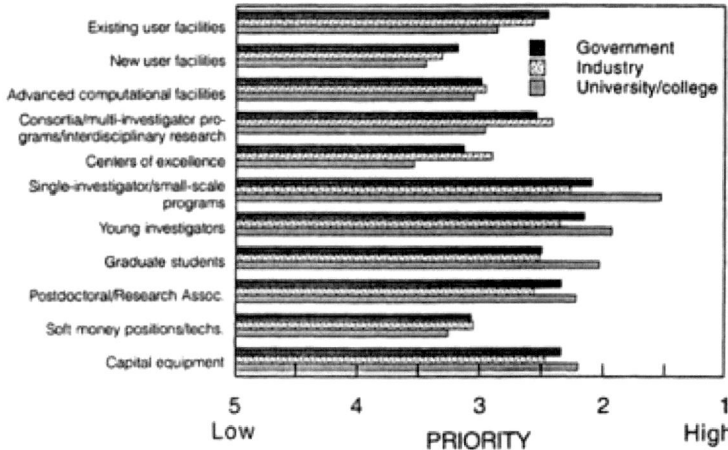

FIGURE D.21 Funding priorities in AMO science. The responses are broken down by employment sector, and the bars represent the mean rating on a scale of 1 to 5 for each category.

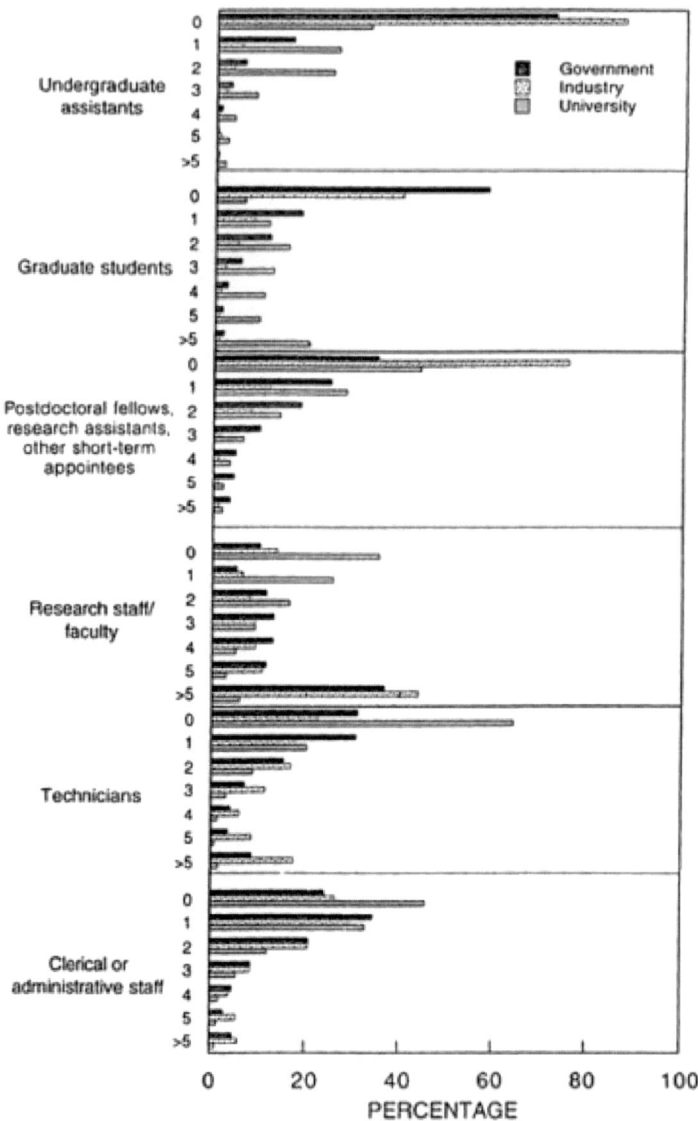

FIGURE D.22 Size of R&D groups or teams in AMO science. The bars represent the percentage of respondents in each employment sector who reported interaction with the number of people indicated in each of the listed categories.

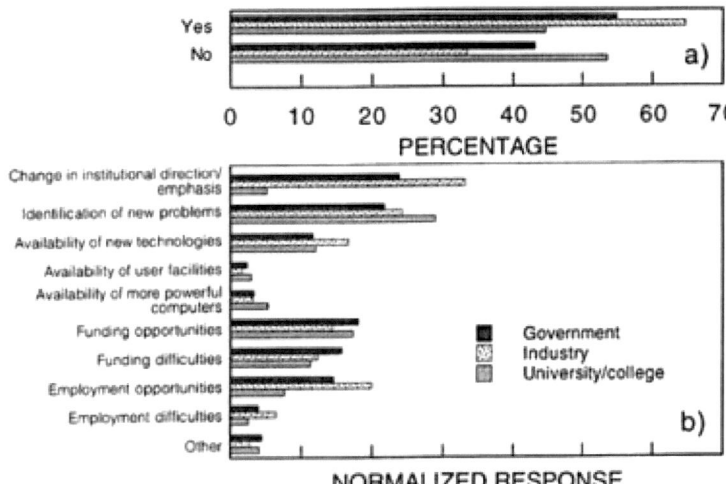

FIGURE D.23 Changes in research direction. (a) The percentage of respondents in each employment sector who reported changing (yes) and not changing (no) their major research emphasis in the past 5 years. (b) The reasons for this change. The data are normalized to the total number of respondents in each sector.

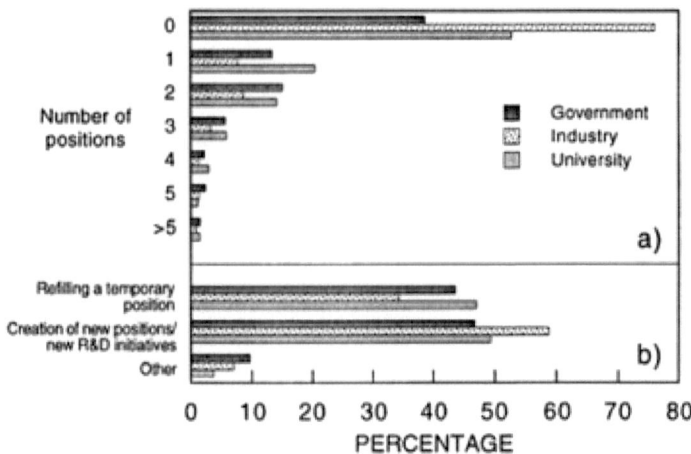

FIGURE D.24 Hiring patterns for temporary positions in AMO science over the past 3 years. The data are broken down by employment sector. (a) The distribution of the number of positions available in each sector. The bars represent the percentage of respondents in each sector who reported that the R&D group or team to which they belong hired or tried to hire the indicated number of AMO scientists to fill a temporary position(s). (b) The reasons these positions became available. The bars represent the percentage of responses in each category.

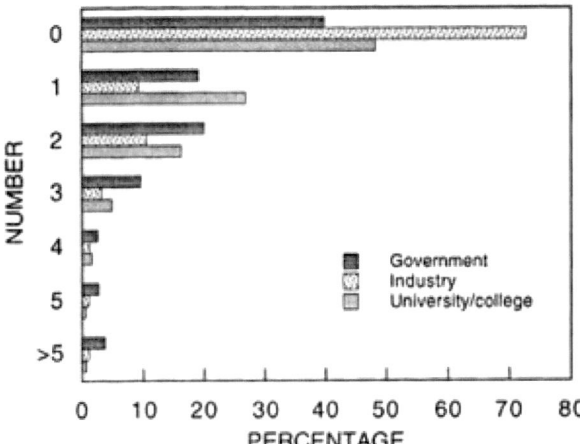

FIGURE D.25a Anticipated recruiting for temporary positions in AMO science. The bars represent the percentage of respondents in each sector who reported that the R&D group or team to which they belong will hire or try to hire the indicated number of AMO scientists to fill temporary positions in the next 3 years.

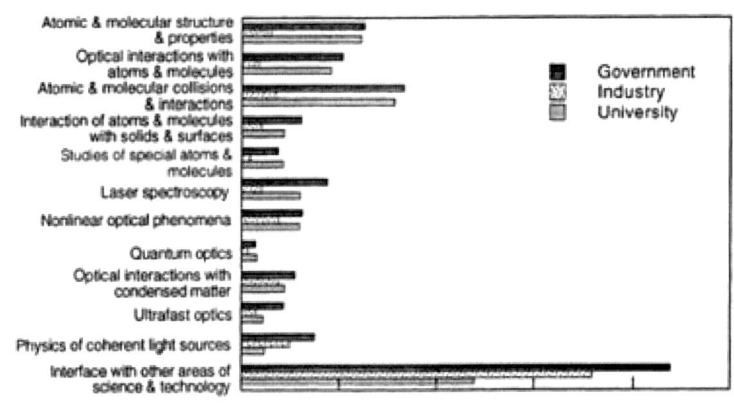

FIGURE D.25b Research specialties of likely candidates for temporary positions. The topics included in each specialty are shown in the Research Specialties Directory (reproduced at the end of this appendix). The data are broken down by employment sector. The bars represent the number of responses in each listed specialty, normalized to the number of respondents in each sector.

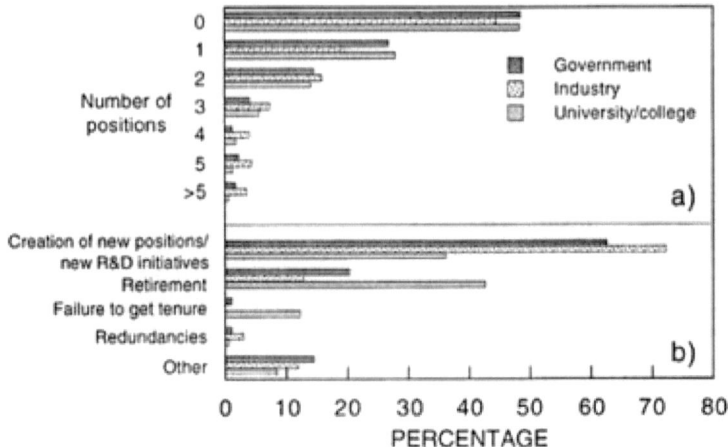

FIGURE D.26 Hiring patterns for permanent positions in AMO science over the past 3 years. The data are broken down by employment sector. (a) The distribution of the number of positions available in each sector. The bars represent the percentage of respondents in each sector who reported that the R&D group or team to which they belong hired or tried to hire the indicated number of AMO scientists to fill a permanent position(s). (b) The reasons these positions became available. The bars represent the percentage of responses in each category.

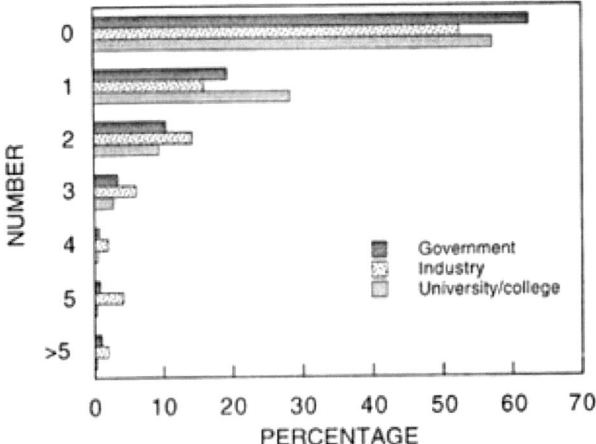

FIGURE D.27a Anticipated recruiting for permanent positions in AMO science. The bars represent the percentage of respondents in each employment sector who reported that the R&D group or team to which they belong will hire or try to hire the indicated number of AMO scientists to fill permanent positions in the next 3 years.

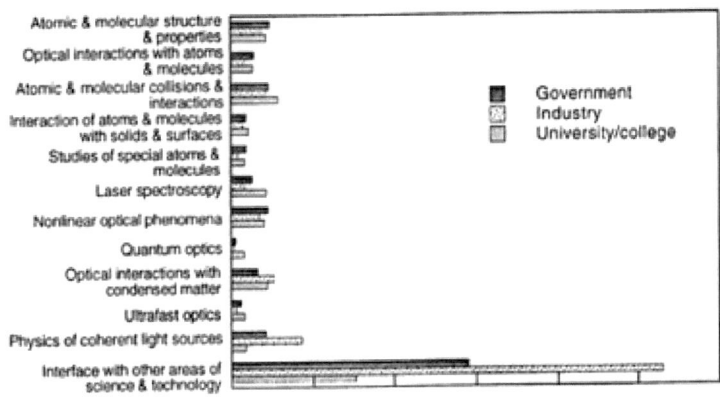

FIGURE D.27b Research specialties of likely candidates for permanent positions. The topics included in each specialty are listed in the Research Specialties Directory reproduced at the end of this appendix. The data are broken down by employment sector. The bars represent the number of responses in each listed specialty, normalized to the number of respondents in each sector.

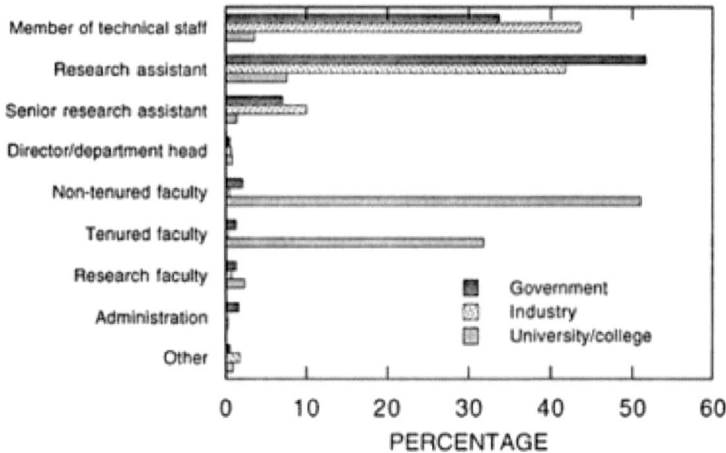

FIGURE D.27c The nature of future permanent positions, broken down by employment sector. The bars represent, for each sector, the percentage of available positions in each category.

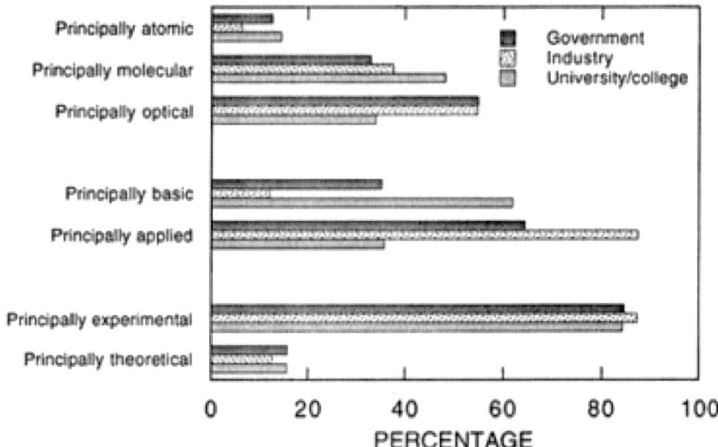

FIGURE D.27d Anticipated principal R&D activities and interests of likely candidates for permanent positions, broken down by employment sector. The bars represent the percentage of likely candidates with the different characteristics indicated.

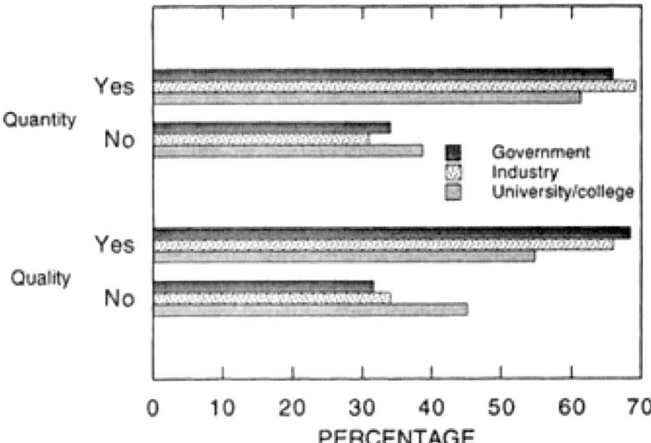

FIGURE D.28 Adequacy of human resources. The bars represent the percentage of respondents in each employment sector who believe that the human resources in their major research area are adequate (yes) or inadequate (no) to meet existing and future needs in terms of quantity and quality.

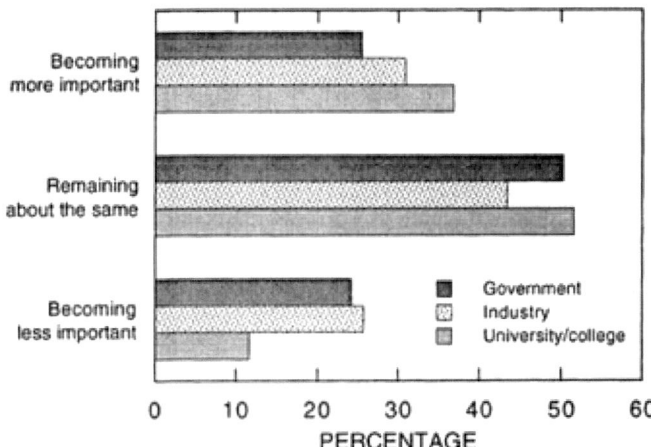

FIGURE D.29 The importance of AMO science. The bars represent the percentage of respondents in each employment sector who believe that AMO science in their organization is becoming more or less important, or remaining about the same.

National Research Council Survey of AMO Scientists · Page 1

EDUCATION AND EMPLOYMENT PROFILE

Name _____
Business Address _____
_____
_____

1. a) What is your highest degree?
   1. ☐ BA/BS
   2. ☐ MA/MS
   3. ☐ Ph.D.

   b) In what year did you receive this degree?

1	9		

2. In what specialty is your highest degree? (CHECK ONLY ONE BOX)
   01. ☐ Atomic and Molecular Physics
   02. ☐ Optical Physics
   03. ☐ Condensed Matter Physics
   04. ☐ Other Physics Specialties _____
   05. ☐ Applied Physics
   06. ☐ Electrical Engineering
   07. ☐ Astronomy and Astrophysics
   08. ☐ Chemistry
   09. ☐ Materials Science
   10. ☐ Other (Specify) _____

3. a) Where do you work? (CHECK ONLY ONE BOX)
   1. ☐ University/College
   2. ☐ Industry
   3. ☐ Federally funded R&D Laboratory
   4. ☐ Non-profit foundation or organization
   5. ☐ Federal Government
   6. ☐ Other (Specify) _____

   b) How many years have you been employed by your present employer?

4. What is your principal position? (CHECK ONLY ONE BOX)
   01. ☐ Member of technical staff
   02. ☐ Research scientist
   03. ☐ Senior research scientist
   04. ☐ Director/Department head
   05. ☐ Temporary appointment (e.g., postdoctoral fellow, research associate...)
   06. ☐ Non-tenured faculty
   07. ☐ Tenured faculty
   08. ☐ Research faculty
   09. ☐ Administration
   10. ☐ Other (Specify) _____

5. To what societies do you belong? (CHECK ALL THAT APPLY)
   01. ☐ AAPT
   02. ☐ AAAS
   03. ☐ ACS
   04. ☐ AGU
   05. ☐ APS
   06. ☐ AVS
   07. ☐ IEEE
   08. ☐ MRS
   09. ☐ OSA
   10. ☐ SPIE
   11. ☐ Other (Specify) _____

6. How would you categorize your major research and development (R&D) activities or interests? (CHECK ONLY ONE BOX IN EACH OF GROUPS A, B AND C)

   A. 1. ☐ Principally atomic
      2. ☐ Principally molecular
      3. ☐ Principally optical

   B. 1. ☐ Principally basic
      2. ☐ Principally applied

   C. 1. ☐ Principally experimental
      2. ☐ Principally theoretical

   If you checked box C2 (principally theoretical), is your work computationally intensive?

   1. ☐ Yes   2. ☐ No

RESEARCH AND DEVELOPMENT ACTIVITIES/ TRENDS

7. Which specialties in the enclosed Research Specialties Directory best describe your R&D activities or interests? *(SELECT UP TO FOUR SPECIALTIES AND ENTER THEIR CODE NUMBERS BELOW)*

   ☐☐ ☐☐ ☐☐ ☐☐

8. Are you involved in collaborative R&D programs with other groups...

   A. Inside the US? 1 ☐ Yes   2 ☐ No
   B. Outside the US? 1 ☐ Yes   2 ☐ No

   IF YOU CHECKED BOXES A1 OR B1:

   C. Are any of these programs cross-disciplinary, i.e., collaborations with disciplines outside AMO science?

   1 ☐ Yes   2 ☐ No

   D. What is the level of your personal involvement in these collaborative programs?

   1 ☐ Very involved
   2 ☐ Somewhat involved
   3 ☐ Peripherally involved

   E. How has your level of personal involvement in collaborative programs changed in the past five years (1987-1991)?

   1 ☐ Increased substantially
   2 ☐ Increased somewhat
   3 ☐ Remained about the same
   4 ☐ Decreased somewhat
   5 ☐ Decreased substantially

9. a) What percentage of your experimental or computational work is conducted at user facilities? ☐☐ %

   b) How has this percentage changed in the past 5 years (1987-1991)?

   1 ☐ Increased substantially
   2 ☐ Increased somewhat
   3 ☐ Remained about the same
   4 ☐ Decreased somewhat
   5 ☐ Decreased substantially

10. a) Which other countries have significant R&D efforts in your major area? *(CHECK ALL THAT APPLY)*

    1 ☐ None   4 ☐ Japan
    2 ☐ France   5 ☐ United Kingdom
    3 ☐ Germany   6 ☐ Other _____

    b) If other countries have significant R&D efforts in your major area, how has the relative U.S. contribution to the total world effort changed in the past 5 years (1987-1991)?

    1 ☐ Increased substantially
    2 ☐ Increased somewhat
    3 ☐ Remained about the same
    4 ☐ Decreased somewhat
    5 ☐ Decreased substantially

11. Using a scale of 1 to 5, how would you rate R&D efforts in your major area in the U.S. compared to those efforts in other countries using each of the following criteria? *(CIRCLE ONE NUMBER FOR EACH ON THE SCALE BELOW)*

	U.S. Very Strong				U.S. Very Weak
A. Innovation	1	2	3	4	5
B. Productivity	1	2	3	4	5
C. Equipment	1	2	3	4	5
D. Facilities	1	2	3	4	5
E. Other	1	2	3	4	5

    (Specify) _____

12. How do you regard the balance between theoretical and experimental work in your major area?

    1 ☐ Too little theoretical work
    2 ☐ About right
    3 ☐ Too little experimental work

RESEARCH SUPPORT

13. What is the source of R&D funding for the group/team to which you belong? *(CHECK ALL THAT APPLY)*

    1 ☐ Federal agencies
    2 ☐ State funds
    3 ☐ Private foundation
    4 ☐ Industrial contract
    5 ☐ Corporate donation
    6 ☐ Internal funds *(IF THIS IS THE ONLY SOURCE GO TO QUESTION 20)*

14. Please give the annual support levels of grants/contracts/donations (in AMO science), including indirect costs, for which you are Principal Investigator (P.I.) or Co-P.I. If Co-P.I. report only your share of the total support. *(IF NONE GO TO 19)*

Source	Annual Support Level ($)
A. Industrial contract	☐ ,000
B. Private foundation (Specify) _____	☐ ,000
C. Corporate donation	☐ ,000
D. State funds	☐ ,000
E. NSF	☐ ,000
F. DOE	☐ ,000
G. DOD	☐ ,000
H. NASA	☐ ,000
I. Other Federal Agency (Specify) _____	☐ ,000
J. Other _____	☐ ,000

15. If you are supported by a Federal mission agency, in which mission category is your research? *(CHECK ONLY ONE BOX)*

1 ☐ Energy
2 ☐ Environment/Health
3 ☐ Space science
4 ☐ Defense
5 ☐ Commerce and technology
6 ☐ Other (Specify) _____

16. What percentage of your total R&D budget is used in salary/stipend support (including both direct and indirect costs) for personnel?

☐ %

17. a) From how many individual grants/contracts/donations (not counting those that amount to ≤ $10,000 per year) do you receive support for AMO science? *(ENTER NUMBER IN THE BOX BELOW)*

☐

b) How has this number changed in the past 5 years (1987-1991)?

1 ☐ Increased substantially
2 ☐ Increased somewhat
3 ☐ Remained about the same
4 ☐ Decreased somewhat
5 ☐ Decreased substantially

18. Taking into account inflation, increased indirect costs, etc., in real terms how has your research funding for AMO science changed in the past 5 years (1987-1991)?

1 ☐ Increased substantially
2 ☐ Increased somewhat
3 ☐ Remained about the same
4 ☐ Decreased somewhat
5 ☐ Decreased substantially

19. Assuming only modest increases in total support for AMO science, do you believe funding for the strongest programs should be increased at the expense of the total number of programs supported?

1 ☐ Yes   2 ☐ No

20. What percentage of your time, on average, do you spend pursuing R&D support for your AMO science?

☐ %

21. Assuming only modest increases in total support for AMO science, on a scale of 1 to 5, how should relative funding priorities be assigned to the following categories in the future? *(CIRCLE ONE NUMBER FOR EACH ON THE SCALE BELOW)*

	High Priority				Low Priority
A. Existing user facilities	1	2	3	4	5
B. New user facilities	1	2	3	4	5
C. Advanced computational facilities	1	2	3	4	5
D. Consortia/multi investigator programs/ interdisciplinary research	1	2	3	4	5
E. Centers of excellence	1	2	3	4	5
F. Single investigator/small-scale programs	1	2	3	4	5
G. Young investigators	1	2	3	4	5
H. Graduate students	1	2	3	4	5
I. Postdocs/research assoc.	1	2	3	4	5
J. Soft money positions/techs	1	2	3	4	5
K. Capital equipment	1	2	3	4	5
L. Other _____	1	2	3	4	5

HUMAN RESOURCES/DEMOGRAPHICS

22. What is the size of the R&D group/team to which you belong, i.e., with how many researchers in each of the following categories do you routinely interact? *(ENTER THE NUMBER, EXCLUDING YOURSELF, IN EACH CATEGORY)*

    A. ☐ Undergraduate assistants
    B. ☐ Graduate students
    C. ☐ Postdoctoral fellows, research assistants, other short term appointees
    D. ☐ Research staff/faculty
    E. ☐ Technicians
    F. ☐ Clerical or administrative staff

23. Have you changed the major emphasis of your R&D activities in the past 5 years (1987-1991)?

    1 ☐ Yes   2 ☐ No

    a) IF YES:
    What were the principal reasons for this? *(CHECK NO MORE THAN THREE BOXES)*

    01 ☐ Change in institutional direction/emphasis
    02 ☐ Identification of new problems
    03 ☐ Availability of new technologies
    04 ☐ Availability of new user facilities
    05 ☐ Availability of more powerful computers
    06 ☐ Funding opportunities
    07 ☐ Funding difficulties
    08 ☐ Employment opportunities
    09 ☐ Employment difficulties
    10 ☐ Other (Specify) _____

24. Has the R&D group/team to which you belong hired or tried to hire AMO scientists to fill one or more TEMPORARY positions with an anticipated duration of two years or less, (e.g., post-doctoral fellow, research associate, etc.) in the past 3 years (1989-1991)?

    1 ☐ Yes   2 ☐ No

IF YES:

a) How many? *(ENTER NUMBER IN THE BOX BELOW)*

☐

b) How did these positions become available? *(CHECK ALL THAT APPLY)*

1 ☐ Refilling a temporary position
2 ☐ Creation of new positions/new R&D initiatives
3 ☐ Other (Specify) _____

25. Do you anticipate that the R&D group/team to which you belong will try to recruit AMO scientists to fill TEMPORARY positions in the next 3 years (1992-1994)?

    1 ☐ Yes   2 ☐ No

IF YES:

(a) How many? *(ENTER THE NUMBER IN THE BOX BELOW)*

☐

b) For the most immediate hires, in which specialties from the enclosed Directory do you feel recruitment would be most likely? *(SELECT UP TO FOUR SPECIALTIES AND ENTER THEIR CODE NUMBERS BELOW)*

Position 1
☐ ☐ ☐ ☐

Position 2
☐ ☐ ☐ ☐

Position 3
☐ ☐ ☐ ☐

26. Has the R&D group/team or academic department of which you are a member hired or tried to hire new staff/faculty to LONG-TERM OR PERMANENT positions with an anticipated duration of greater than two years in AMO science in the past 3 years (1989-1991)?

    1 ☐ Yes   2 ☐ No

IF YES:

a) How many? *(ENTER THE NUMBER IN THE BOX BELOW)*

☐

Page 5

b) How did these positions become available? (CHECK ALL THAT APPLY)

1. ☐ Creation of new positions/new R&D initiatives
2. ☐ Retirement
3. ☐ Failure to get tenure
4. ☐ Redundancies
5. ☐ Other (Specify) _____

27. Do you anticipate that your group/academic department will try to hire new staff/faculty to LONG-TERM OR PERMANENT positions in AMO science in the next 3 years (1992-1994)?

1 ☐ Yes   2 ☐ No

IF YES:

a) How many? (PLEASE ENTER THE NUMBER IN THE BOX)

[   ]

b) For the most immediate hires, in which specialties from the attached Directory do you feel recruitment will be most likely? (SELECT UP TO FOUR SPECIALTIES AND ENTER THEIR CODE NUMBERS BELOW)

Position 1
[ ][ ][ ][ ]

Position 2
[ ][ ][ ][ ]

Position 3
[ ][ ][ ][ ]

c) What will be the type of these appointments? (ENTER NUMBER FROM LIST BELOW)

Position 1 [ ]
Position 2 [ ]
Position 3 [ ]

1. Member of technical staff
2. Research scientist
3. Senior research scientist
4. Director/department head
5. Non-tenured faculty
6. Tenured faculty
7. Research faculty
8. Administration
9. Other (Specify) _____

d) How would you characterize the likely R&D interests of the person(s) recruited? (CHECK ONLY ONE BOX IN EACH OF GROUPS A, B AND C FOR EACH POSITION)

		Position 1	Position 2	Position 3	
A.	1	☐	☐	☐	Principally atomic
	2	☐	☐	☐	Principally molecular
	3	☐	☐	☐	Principally optical
B.	1	☐	☐	☐	Principally basic
	2	☐	☐	☐	Principally applied
C.	1	☐	☐	☐	Principally experimental
	2	☐	☐	☐	Principally theoretical

e) If you checked box(es) C2 (principally theoretical), would you expect to hire someone whose work is computationally intensive?

1 ☐ Yes   2 ☐ No

28. Do you believe human resources in your major area are adequate to meet existing and future needs in terms of...

A. Quantity?   1 ☐ Yes   2 ☐ No
B. Quality?   1 ☐ Yes   2 ☐ No

29. On balance, how is the overall importance of AMO science in your organization/institution changing?

☐ Becoming more important
☐ Remaining about the same
☐ Becoming less important

COMMENTS: _____

_____

_____

Thank you very much for your participation. Please return the completed questionnaire in the enclosed postage-paid envelope to:

AMO Survey
Board on Physics and Astronomy
National Research Council
2101 Constitution Avenue
Washington, D.C. 20418